# COMPUTERS IN FORESTRY

# COMPUTERS IN FORESTRY

edited by
W.L. MASON AND R. MUETZELFELDT

Proceedings of a Conference on the
Application of Computers to the Management and Administration of Forests,
the Harvesting and Marketing of Timber
and to Forestry Research

Heriot-Watt University, Edinburgh, UK
11-14 December, 1984

INSTITUTE OF CHARTERED FORESTERS

Institute of Chartered Foresters
22 Walker Street, Edinburgh

Typeset by Waverley Graphics Limited
New Street, Edinburgh.

Printed in Great Britain by
Redwood Burn Limited
Trowbridge, Wiltshire

British Library Cataloguing
in Publication Data

Computers in Forestry:
proceedings of a conference held at
Heriot-Watt University, Edinburgh,
11-14 December 1984.
1. Forests and forestry—Date processing
I. Mason, W.L. II. Muetzelfeldt, R.I.
634.9′028′5 SD381.5

ISBN 0 907284 07 8

## CONTENTS

## PART IV. APPLICATIONS IN MAPPING AND HARVESTING



## PART V. COMPUTERS IN THE WOOD PROCESSING INDUSTRIES

## PART VI. DEVELOPMENTS IN RESEARCH AND MANAGEMENT



## PART VII. SUMMING-UP

# PREFACE

The decision of the Council of the Institute to host an international conference on the use of computers in forestry followed a suggestion from two of its active younger members, Mr Mark Pritchard and Mr Alan Stevenson. These two, joined by Mr Guy Watt, provided much of the driving force which organised and ran the meeting. The detailed organisation was provided by the Institute's Secretary and her Assistant, Mrs Margaret Dick and Miss Sally Bryce. To all of these the Institute is deeply indebted for the excellence of the meeting which participants clearly both enjoyed and valued for its technical content.

Finally the Institute thanks the joint editors of the conference proceedings, Mr Bill Mason and Dr Robert Muetzelfeldt, for the immense task of preparing the material for publication, work which began long before the meeting itself and has continued long after.

The timing of the conference was particularly effective in relation to the decisions being taken by the Forestry Commission and several forestry consultants and companies in specifying the place of computers in the management of forest resources in the UK. It appears to be a very proper role for a Chartered Institute that it should lead professional opinion in such development, in this instance laying out the technical possibilities and enabling the leaders of forestry in the UK, not only to see the state of the art in computing but also to discuss the real possibilities and problems of its introduction with professionals directly involved from the rest of Europe and North America. The rewards of those involved in the organisation of the conference were the expression of appreciation from participants and the consequent belief that the purpose had been achieved. The organisers and the Council of the Institute are sincerely grateful to the speakers and the commercial firms who contributed papers and demonstrations which were the substance of this very successful meeting.

The conference was supported by trade displays of computer hardware and software organised by Forestry and British Timber. Details of exhibitors may be found in the December 1984 and February 1985 issues of that journal.

W.E.S. MUTCH

# PART I

# INTRODUCING COMPUTERS TO A COMPANY

# COMPUTER HARDWARE AND SOFTWARE

F. STACEY

*Department of Computer Science, University of Edinburgh*

SUMMARY

A brief description is given of the structure and operation of a digital computer. Emphasis is placed on those topics which an inexperienced user is most likely to encounter.

INTRODUCTION

This paper aims to introduce the structure and mode of operation of a digital computer to people with little or no background in the subject. Sufficient context is provided to permit the explanation of a number of frequently heard jargon terms. Computing machinery is considered first, followed by the programs which exploit its potential.

COMPUTING MACHINERY—HARDWARE

A fundamental characteristic of a digital computer is the technique by which it represents the data which it processes. Any digital device is ultimately constructed by interconnecting a large number of very simple devices, each of which can exist in only 2 different states. Figure 1 shows a single 2-state storage device. Its inputs and

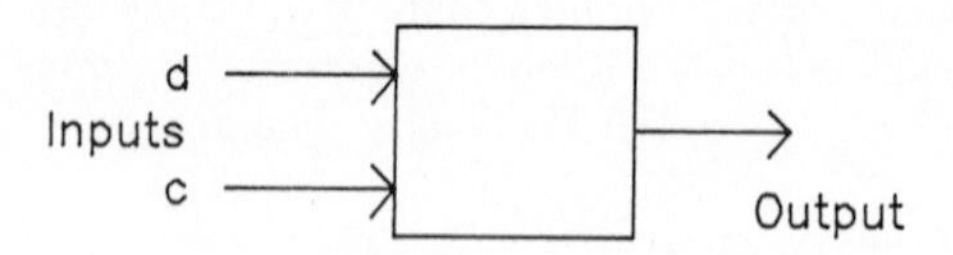


FIGURE 1. 2-state storage device.

outputs are simply wires which communicate the state of one device to another. The states are often denoted by digits 0 and 1, but they have no numerical significance in this context. Such a device can be made to change (instantaneously) to a given state by the appearance of certain combinations of states on its input lines, other input

TABLE 1

| Inputs c | d | Meaning | Output |
|---|---|---|---|
| 0 | 0 | ignore | Unchanged |
| 0 | 1 | ignore | Unchanged |
| 1 | 0 | Go into state 0 | 0 |
| 1 | 1 | Go into state 1 | 1 |

TABLE 2

| 3 bit row | | | Possible Data Represented | | |
|---|---|---|---|---|---|
| 1 | 2 | 3 | Colours | Trees | Loudness |
| 0 | 0 | 0 | Black | Oak | ppp |
| 0 | 0 | 1 | Blue | Ash | pp |
| 0 | 1 | 0 | Green | Elm | p |
| 0 | 1 | 1 | Yellow | Beech | mp |
| 1 | 0 | 0 | Red | Sycamore | mf |
| 1 | 0 | 1 | Magenta | Lime | f |
| 1 | 1 | 0 | Brown | Chestnut | ff |
| 1 | 1 | 1 | White | Maple | fff |

states being ignored (Table 1). A 2-state device stores 1 *bit* of data.

A computer represents the data it is to process by rows of bits. Table 2 shows how a row of three bits is capable of representing any one of the 8 different values for the three types of data suggested. The more bits that are available the wider the range of values that can be represented. A modern computer has a store of many millions of bits ready for immediate processing.

Its storage capacity is an important measure of the power of a computer. This is usually expressed in terms of *bytes* which are simply rows of 8 bits. Even a byte is a miniscule unit so the *Kilobyte*, 1024 bytes or the *Megabyte*, 1,048,576 bytes, are used. These peculiar numbers, close to a thousand and a million, are derived from the number of distinct bit patterns that can be produced by a row of 10 and 20 bits respectively.

A byte can represent 256 distinct values, which is a little more than the number of different printing characters used in typical textual information. Computers, therefore, use bytes to represent printed characters. On the basis of 1 character per byte a long paragraph occupies a Kilobyte, a 200 page paperback half a Megabyte and the complete Oxford English Dictionary would occupy approximately 300 Megabytes.

*Integrated circuit (silicon chip) technology*

Thirty years ago a 1 bit store occupied approximately 1 cubic inch. With today's technology of imprinting the devices on the surface of a silicon crystal (chip), as many as 100,000 bits may occupy the same cubic inch. The cost of that cubic inch has increased comparatively little in real terms over that period. It currently stands at between 10 and 100 bits/penny. This enormous change in density and the reduction in cost are the prime factors which have brought the computer into almost every work-place and many homes of the developed world.

*The anatomy of a computer*

Although each consists primarily of 2-state devices, there are a number of distinct units in any computer. Figure 2 shows the principal functional units of a conventional computer and their inter-connection. Each will be described briefly in turn.

*The Processor*. This is the principal active component. It generally copies data from other units, processes these to produce new results which it can send back. The data are always in the form of rows of bits representing, perhaps, numbers which the

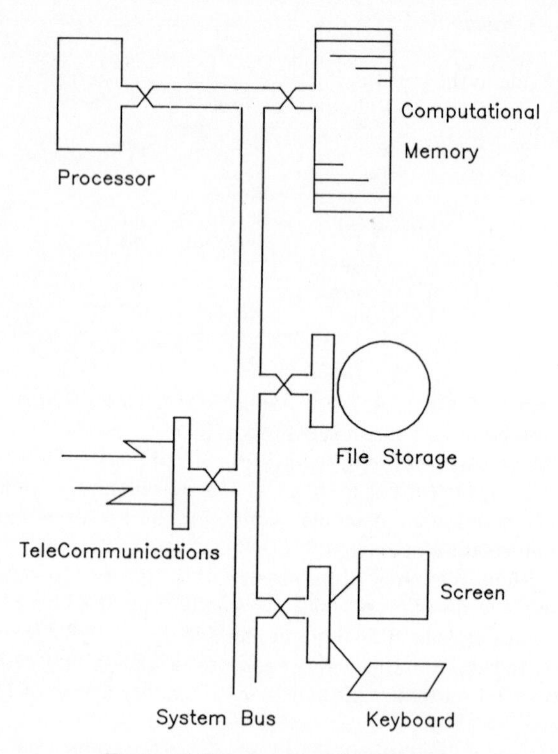


FIGURE 2. The anatomy of a computer.

processor can add, subtract, compare, etc. to produce new numbers. A processor has a repertoire of a few hundred operations which it can perform on data from any part of the system. The sequence of operations which it follows is determined by a *program* of these *instructions*. Each instruction is itself represented by a row of bits in the computational memory.

The power of a processor, that is how fast it can perform useful work, is determined by many factors of which some important ones are:

a) Its speed. This is measured in millions of instructions per second (MIPS) which it can process. Available processors range from 0.5Mips for a home computer to 100Mips for the most powerful machines.

b) Its wordsize. Each processor model deals principally with rows of bits of some fixed length. Wordsize currently available are 8, 16 and 32 bits. The bigger the wordsize the bigger the unit that can be processed in one operation and hence the faster the processor can perform a given job.

c) Instruction Set (Order Code). The limited repertoire of instructions known to the processor is designed to enable a wide variety of commonly desired effects to be achieved with the minimum number of instructions.

A *micro-processor* is one which is fabricated on a single silicon chip. Though smaller and cheaper than its multi-chip counterpart it is functionally the same. Currently few micro-processors are genuinely 32 bit, those built into micro-computers are mostly 8-bit, with a few more expensive models incorporating 16 bit processors. A *micro-computer* is simply a computer built round a micro-processor.

*Computational Memory.* This is the store in which both the data and the program immediately available to the processor reside. All computational memory is strictly *random access* memory, meaning that every byte in it is equally readily accessible to the processor. The abbreviation RAM is used in a slightly more restricted sense to refer access memory which can be both read (copied from) and written to, and which is volatile in that its contents are destroyed when power is switched off. In contrast to this, ROM is read-only computational memory, whose contents are fixed during its manufacture and remain intact when power is switched off. Essential programs are placed in ROM, especially in cheap machines with no fast reliable permanent storage medium.

*The File Store.* Computational memory is fast and expensive and usually transient. Disc storage is used for permanent storage of data and programs in all but the cheapest systems. Its costs are about 1000 bits per penny, although it is also a 1000 times slower than computational memory. The disc is a non-magnetic circular plate on whose surface is a thin coating of magnetic material. Data are recorded in concentric circular tracks on this surface by a technique which is a development of audio and video tape recording. The disc spins while a single read/write head can move radially to the desired track. A randomly chosen area can be accessed in approximately 1/100th of a second.

*Floppy* discs are the cheapest, based on a flexible plastic disc of 3.5 or 5 inches diameter, each with a storage capacity of approximately 1 Megabyte. *Winchester* discs are more expensive, use rigid discs of various sizes from less than 5 to 14 inches in diameter and achieve the highest possible recording densities. Multi-disc spindles having capacities ranging up to 500 Megabytes are available.

*The User Terminal.* This is the means by which the user communicates with the machine. The ubiquitous device for submitting data to the machine is the tyepwriter (qwerty) keyboard. Output is usually presented via a television screen. In the case of a visual display unit (VDU), only text will be shown (and precious little of that) although, as memory prices have fallen, it has become economical to include graphics output. A monochrome picture can be represented as an array of bits each representing one spot of a closely packed array of spots or *pixels* on the picture. To give a high fidelity image of one page of, say, a typical text book requires approximately 1 Megabyte if the picture is monochrome but several times that if it is coloured.

*Telecommunication.* Improvements in this technology have allowed computers to communicate reliably over large distances. Thus it has become common to set up *networks* of inter-connected computers. Such networks are divided into 2 broad categories. *Local area networks* (LANs) are confined to distances of a few kilometres, usually a building or a site. The links employed can operate reliably at speeds of up to 1 Megabyte $sec^{-1}$. In contrast, *wide area networks* span hundreds of miles. National and inter-national telecommunication facilities must be used which currently limit reliable operation to speeds of several Kilobytes $sec^{-1}$.

*Types of computer*

Available machines fall into a few broad categories. *Personal* computers are generally micro-computers of a size, power and cost such that one person can afford to use it exclusively. *Home* computers are at the lower end of this range with office computers at the upper. The more powerful machines are referred to as *mainframes*.

These, often hidden away in remote computer rooms, are used concurrently by many people since no one person could profitably keep it occupied for very long. While one person is pausing for thought the machine can (and usually does) execute several other peoples' programs. The most powerful of all are the *super computers,* running at 100Mips and more which are used for (amongst other things) the accurate prediction of the weather. A third, intermediate, category of machine, the *mini-computer,* is still sometimes referred to. It is a reflection of the rate of change of computer technology that scarcely any manufacturer currently produces a model referred to in this way. The low end of the market has been taken over by the personal machine and manufacturers have been forced to improve their price/performance to such an extent that the upper end of the mini-computer range has moved into the mainframe category.

### COMPUTER PROGRAMS—SOFTWARE

The impressive collection of hardware is ineffectual without programs to drive it. To create a program, the programmer must use a *programming language.* This a written language with very rigid rules of syntax. When a program has been written it can be read and obeyed by a computer, although often the computer must first translate the written version into *machine code* (rows of bits) and then obey that.

An example program fragment is given below, which goes through data on each of a thousand trees in turn. The data include the type of the tree and the year in which it is expected to be felled. From this the program computes the number of oak trees to be felled in the period up to 1990. The language used is similar to BASIC.

```
oaks this decade=0
FOR tree=1 TO 1000
   IF (type (tree)=oak) AND (estimated fell year (tree)<=1990) THEN oaks
      this decade=oaks this decade + 1
NEXT tree
PRINT "Estimated oaks to end of decade=";
PRINT oaks this decade
END.
```

#### *The operating system*

This collection of programs has overall control of the operation of the machine. Its functions include providing a *filing system* which makes file storage convenient. Files are identified by names of their owner's choice, and are allocated space on the disc as and when needed without the programmer being concerned with where they are. All but the crudest operating systems allow several programs to co-exist in the computer's memory: the processor works on each in turn in rapid succession. To the outside observer it appears as if such a *multi-programmed* machine is performing tasks simultaneously. This technique can make very effective use of a fast processor.

The remit of the operating system is often large enough to include specifying the sequence of keystrokes required to initiate or control overall actions such as: listing names of known files, making a copy of a file on a printer or starting up the financial planning program. Different models of computer may have the same operating system, in which case a person familiar with the use of one model should easily be able to change to another. The operating systems commonly available include CP/M, PC-

| | |
|---|---|
| *Mail:* | Compose ack |
| *To:* | A. Smith, B. Brown, J. Bloggs |
| *Subject:* | Annual Reports |
| *Acknowledge:* | P. Perkins |
| *Text:* | Final reminder everyone! |
| *:* | Deadline only 2 weeks off |
| *:* | : |
| *Send Now?* | No |
| *After:* | 31st December, 1984 |

User types plain text
*Computer prompts in italics*

FIGURE 3. Electronic mail

DOS, MS-DOS, UNIX and several others.

*Programming languages*
There are many different programming languages, but they fall into a few broad categories.

*Assembly languages.* These are the crudest programming languages which simply represent the machine-code instructions in a symbolic textual form. For example, simple verbs represent arithmetic operations such as ADD, SUBTRACT, and so on.

*High Level languages.* These present the machine in a very much more abstract way. This allows the programmer to think in a manner more appropriate to the problem to hand than to the machine which will solve it. There are many high level languages including:

Cobol and Fortran are showing their age since they were designed 25 years ago, but they are still much used in the business and scientific worlds respectively.

Pascal and C are modern languages which cover broadly the same types of problem as Cobol and Fortran.

PROLOG, a completely new type of programming language whose supporters claim it could revolutionise the problems of writing computer programs.

Another generation of programming languages (some say the 4th generation) is becoming available, which allows the programmer to use very much more natural modes of expression. These are specially prevalent in the area of *data base* query languages, that is languages which enable complex questions to be asked about a large set of inter-related data.

*Application packages*
This term is used to describe a wide variety of useful facilities ultimately provided by the machine. Some are merely extensions to existing languages, requiring their users to be familiar with the mother language. Others are complete languages in themselves, usually providing the user with a helpful and readily understandable environment in which to work. Examples of the latter are *word processing* systems for the preparation, printing and filing of high quality documents, and *electronic mail* systems for the reliable transmission of personal messages between members of a community of users. Figure 3 shows a sample of the preparation of a piece of mail, to be sent to 3 people, but not before the given date. As each recipient accepts the mail

the mailer will send an acknowledgement to, say, the sender's secretary.

## CONCLUSION

Despite (or perhaps because of) the endeavours of computer professionals it is often very difficult to make efficient use of computers without some knowledge of their inner workings. This paper has briefly surveyed some of the concepts which are relevant in this respect. The author is convinced that such knowledge will remain essential for some considerable time to come.

# COMPUTER SELECTION WITH MINIMAL RISK

A. McKELLAR

*Principal Consultant, Arthur Young, George House, 50 George Square, Glasgow G21 1RR*

SUMMARY

Purchasing a computer system can be a harrowing experience. Before computers can be introduced into an organisation, four procedures should be carried out. These are, in turn, strategy determination, computer selection, system implementation and performance monitoring. Correct use of these procedures will minimise the risks involved in complete selection and ensure that introduction of a computer enhances business efficiency.

INTRODUCTION

The opening line in *The Tale of Two Cities* reads—'it was the best of times, it was the worst of times'.

Paradoxically, that is the state of the computer industry today. On the best side, there are machines to suit every pocket, containing practically every function imaginable. On the worst side are the decisions such as 'should I buy, what do I buy, and what do I need to know before I buy'?

Few successful businessmen would embark on the production of an entirely new range of products without first assessing the market, deciding on the level of investment and the likely return, planning the marketing opportunities and analysing the sales potential.

It should be exactly the same with computers. In some cases they are purchased in order to carry out a single task more quickly and efficiently than the existing procedure. For example, using a micro-computer to register the contents of a warehouse rather than adopt the now-outdated method of entering by hand all the items in bulky register books. The same outline can be applied to office accounting systems or for the payment of salaries.

A more complex situation arises when owners recognise the value of the data relating to their business and the information which can be derived from that data. At that point a computer becomes more than simply an aid and is used as a practical tool in highlighting exceptions. It also allows accurate information to be accessed instantly so allowing time for decisions to be made based on that information. In many cases it may assist in identifying an alternative approach.

However, before introducing a computer into a business there are four clearly defined stages which should be carried out:

Strategy determination
Selection
Implementation
Performance monitoring

Because each stage is not only important in itself, but affects, and is affected by, all the others, it is imperative to discuss each independently to obtain a more detailed understanding of what each stage involves.

## STRATEGY DETERMINATION

### *Information Strategy*

The company must examine the data and flow of data which are needed to support all of its systems, manual and computer-based. The data which will be required to meet the needs of the business strategy are also identified. From this, the interrelationship between groups of data will emerge. This information will be used to identify the computer systems needed to meet the business objectives. The company must examine its plans and forecasts for the future and recognise their implication.

The next step is to use all the information gathered so far to develop a data processing strategy.

### *Data Processing Strategy*

The requirements must now be established, that is the required computer systems, and the necessary hardware, software, people, accommodation and training.

The desired timetable for development of the system must be decided and an assessment of the costs and likely benefits should be made.

Having established the strategy, the next stage is selection.

## SELECTION

The approach we recommend is designed to enable you to sign a contract with performance guarantees. This simply means that the responsibility for proposing technical solutions to commercial problems is that of the supplier, who undertakes to provide the hardware, software and implementation support you need, and to leave an operational system guaranteed to perform its function. By adopting this approach, you can avoid having to make technical comparisons, relying instead on a comparison of prices, guarantees, levels of service offered, security, reliability of the suppliers, and so on—commercial issues on which you can pass judgment.

Such a form of contract is not readily offered; you will have to impose constraints on the process of tendering in order to get it. A well-tried standard procedure makes this possible:

Define your requirements
Issue an invitation to tender
Evaluate the responses
Agree the contract

### *Define Your Requirements*

This takes time and effort but is a necessary foundation for a good decision. Unless it is done well, what follows will have little chance of success. This preliminary step will determine where a computer will be most likely to produce benefits.

The next step is to define the required applications in terms of:

Frequency and content of regular reports
Types of transaction
Volume of transactions

Size and contents of permanent data files, e.g. inventory files, customer files, supplier files

Nature and frequency of any occasional requirements for information

Control requirements

This information must then be presented in a form that suppliers will recognise. It usually takes the shape of a document known as a Statement of Requirements, the typical contents of which would be:

*Introduction*

This section briefly describes your company, the nature of its business and its previous experience of data processing.

*Commercial constraints*

This section describes the commercial constraints within which the systems need to operate. Factors taken into account should include, for example, geography and seasonal fluctuations in business.

*System requirements*

This section describes the required systems and is subdivided into: system flowchart, reports to be produced, input to the system, brief processing description, data volumes.

Volumes should make allowance for expected business expansion over a three-year period.

This information should be documented and will provide the basis for the next stage.

*Issue an invitation to tender*

You then need to decide whom you are going to ask to tender for the supply of your system, and this can be a difficult decision. There are literally hundreds of suppliers of small business systems with a wide range of products. Some are primarily suppliers of minicomputer hardware who sell little application software and rely instead on other firms to develop and market software for their equipment. Other suppliers sell both hardware and software as a package.

As some firms are unwilling to take the trouble to respond when there are many tenders, it is better to draw up a shortlist of suppliers, usually three or four. In selecting them it is important to identify those with an adequate software package or demonstration experience in the application being considered, and with local personnel who are able to modify software if necessary and maintain equipment.

The document requesting quotations for a computer and the necessary programs is called the Invitation to Tender, and usually consists of:

Your schedule for getting the chosen system into operation

The contractual terms you seek

The format you wish the tender to take

Your Statement of Requirements

It should be issued to the shortlist of suppliers, the date for tenders to be returned, allowing time for them to prepare a proper document. About four weeks is usual; any longer and suppliers tend to put it to one side as non-urgent, and it inevitably gets forgotten.

You should talk to each supplier about your requirements and stress those which you consider most important. Impress upon suppliers that it is your intention to buy a complete service, including support during the design and installation stages. Give them an opportunity to ask questions and allow them to suggest modifications to your processing requirements. A supplier knows best how his machine operates and may be able to offer constructive advice. Do not, however, accept fundamental changes to the Statement of Requirements.

*Evaluate the responses*

To decide which supplier has made the best offer, you must first determine your criteria for selection. There is no hard-and-fast rule about this as each organisation needs to decide for itself what it considers important. Criteria might, for example, include:

Financial viability of suppliers
Cost of installing the machine (including hardware, software and installation)
Monthly operating costs
Ability of the supplier to meet the schedule
Capability for expansion of the solution proposed
Reliability and security
Delivery
Environmental requirements for the machine
Support and training offered during and after installation
Arrangements for the maintenance of hardware
Location of supplier

It is worthwhile being thorough in evaluating tenders. They should be read carefully and the suppliers' promises extracted and recorded against the list of criteria for comparison.

*Agree the contract*

The installation of almost every computer system is accompanied by a tremendous amount of frustration. Delays, equipment problems, programming mistakes and changes will almost always occur. Most of these problems will be solved by compromise.

In order for the compromise to work best for you, you must have a position of strength—and this can only be provided by the words in your contract. You may need to have two separate contracts, one for system development, the other for hardware maintenance. Usually it is the system development contract that causes negotiating problems.

Of course, if it is necessary to go to court to solve a problem, the contracts will be important and you should be protected. But, as a practical matter, contracts should be designed to avoid litigation and to provide a basis of agreement for procedures involved in installation and conversion. You will probably need professional advice on what parts of the supplier's standard contract to accept and what additional clauses are necessary.

With a good contract, installation will be easier, but be prepared for the worst. Recognise the importance of the contract and how it forces all parties to plan and agree on objectives.

IMPLEMENTATION

This is fairly straightforward but remember the importance of the proper selection of your own staff, and the correct type of training for them. If these have not been carried out at this point then there may be problems at a later stage. Assuming that this area has been covered, then implementation should not cause any problems.

*Planning*

The need for planning is self-evident but should ensure no task is missed out and that adequate margins for errors, problems and delays are built into timetables. Each activity, for example, electrical supply, data input, etc., should be listed by time and responsibility in order that effective progress monitoring can take place.

*Documentation*

This is probably the most neglected area in the implementation of a computer system. The life expectancy of software application systems has now risen to ten years and above. During that period the system must be maintained, supported and enhanced—probably by many different people. Efficient documentation is vital since users must have a clear reference governing what they do in their day to day business lives to the computer.

*Data Access Security*

The risks which exist for all companies using computers must be identified and assessed in terms of potential damage to the company and its business. In addition, for computers storing personal data about employees or members of the public, the requirements of the 1984 Data Protection Act must be observed.

Briefly the terms of the Act are to protect the public from misuse of information about them held in computers. The legislation permits any individual in the UK to find out who is holding personal information on computer about him, or her, the details of the information and why it is being stored. The Act makes it mandatory for any company compiling that type of information to register that they are engaged in that practice with the Home Secretary or his nominees. This of course leads into the field of computer security and the safeguards which should be taken to provide protection against breakdowns or unauthorised access to computers storing personal data or records.

*Physical Security*

This is a most obvious area of concern but one which is often not fully understood. It is common to find physical access to a machine barred by electronic locks. Fire precautions are extensive in most cases. Terminals are protected by locks or passwords, or sometimes both. Areas of danger may be under-training of operators resulting in expensive errors. Other problem areas are in the lack of back-up information or that back-up tapes or discs may be held in the same building as the machine. These should be ideally stored separately to avoid disastrous loss of information.

*Software Installation*

Installation of a software package involves careful appraisal of those parts of the system which require modification to suit company requirements. All relevant documentation must also be changed at this point.

*Training*
Both users of computers and the computer staff must be adequately trained before implementation takes place. This point has been fairly well covered earlier but the importance of the part it plays cannot be stressed too often.

Having properly followed all the above procedures the company should now find that it will not be the victim of the 'It'll be all right on the night' syndrome and that their newly acquired system will work well.

## PERFORMANCE MONITORING

There are occasions when, at this stage, the original objectives set out in the strategy planning stage have altered. This is not common but it will do no harm whatsoever simply to refer back to the original business objectives defined some time ago. Do they still apply? If not, do the changes, or alterations make significant differences to the computer system? These points should be assessed at this stage.

Obviously you must ensure that the systems meet your stated business requirements.

## CONCLUSION

The age of information technology is firmly established worldwide. More and more we are relying on that technology to assist us in business and, indeed, in our personal lives. Therefore it is vitally important that all the proper steps are taken before investing in that technology.

If the procedures outlined above are adopted then you will have minimised the danger areas and will be in a position to make the most of the properly selected computer, at the right price, designed to carry out specific functions to enhance your business efficiency.

Most computer hardware installed will be out of date within one year, due to developments in technology. However, first time users should avoid pioneering new systems and accept that they are buying a business tool which, if selected properly, will provide business benefits whether or not it is 'next year's model'.

# USER DRIVEN DEVELOPMENT OF A DECISION SUPPORT SYSTEM FOR FORESTRY PRODUCTION CONTROL

E. TORSTEN LUNDQUIST

*Infologigruppen AB, Stockholm, Sweden*

SUMMARY

An organisation is a human activity system defined as a set of cooperating human beings. It is much more complicated than a technical system and does not have well-defined goals. Can such a system be modelled?

We have to deal with a difficult situation where people continuously change their views in a never-ending learning process. This process is itself desirable.

The history of computer installations in information systems is not very encouraging, for two reasons. First, we try to use the same 'hard' methods in a human activity system as we have used for our computer programs. Second, we have up to now used conventional programming tools for computer program development. Both will, sooner or later, always lead to an 'unbridgeable gap' between the computer programs and the users' needs.

In this paper we give a very brief description of a model of an information system for control of forestry production. In developing this system we have discovered some problems which we will discuss. The emergence of 'fourth generation languages' can suggest solutions to many of these problems. This not only permits the system designer to produce computer programs faster, but, of greater importance, opens up the possibility for non-programmers to obtain real control, real 'user-driven computing'.

## INTRODUCTION

A computer is an example of a system which is characterised by its predefined behaviour, with the same input always leading to the same output. It is an example of a relatively primitive system which, in a hierarchy of systems of increasing complexity, is ranked nearest to such simple structures as bridges (Checkland, 1981, p105).

On a much higher level are the socio-cultural systems we call organisations, where human beings cooperate in pursuit of agreed goals.

Computers and computer programs can be developed using well-known methods. Organisations, on the other hand, cannot be successfully developed by these methods. This creates problems when we install computers in our organisation, and try to represent the latter by the same methods we used when we developed the computer programs.

The purpose of an organisation (Figure 1) is, as for a computer, to transform some kind of input into some kind of output. In order to do this, the people in the organisation communicate and help each other with physical actions. They interfere with and control each other's decisions often via data. These data are incorporated

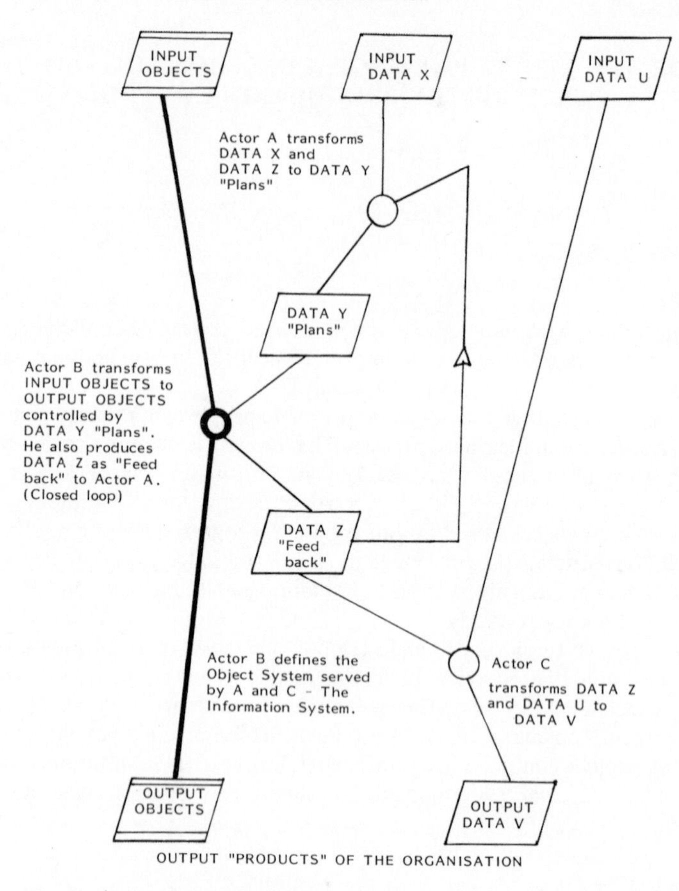
INPUT TO THE ORGANIZATION
INPUT OBJECTS
INPUT DATA X
INPUT DATA U
Actor A transforms DATA X and DATA Z to DATA Y "Plans"
DATA Y "Plans"
Actor B transforms INPUT OBJECTS to OUTPUT OBJECTS controlled by DATA Y "Plans". He also produces DATA Z as "Feed back" to Actor A. (Closed loop)
DATA Z "Feed back"
Actor B defines the Object System served by A and C - The Information System.
Actor C
transforms DATA Z and DATA U to DATA V
OUTPUT OBJECTS
OUTPUT DATA V
OUTPUT "PRODUCTS" OF THE ORGANISATION

FIGURE 1. The organisation—a 'soft' human activity system.

into the recipient's frame of reference and change his actual opinions in a way which depends on his interpretation, the assumptions he makes and his basic view. This change defines the information content of data.

It is the information content of the data and not the set itself which controls what decision is made and what act is going to be undertaken. The selection of a set of data to use in order to control a human is (happily) rather problematic.

*Development of Information Systems*

Given the aim of developing a computer-based information system to help with decision-making, we are then faced with a number of questions:

Which people make which decisions?

What data are 'best' in different decision situations?

How are these data best presented?

It follows from the previous discussion about information that the answer to these

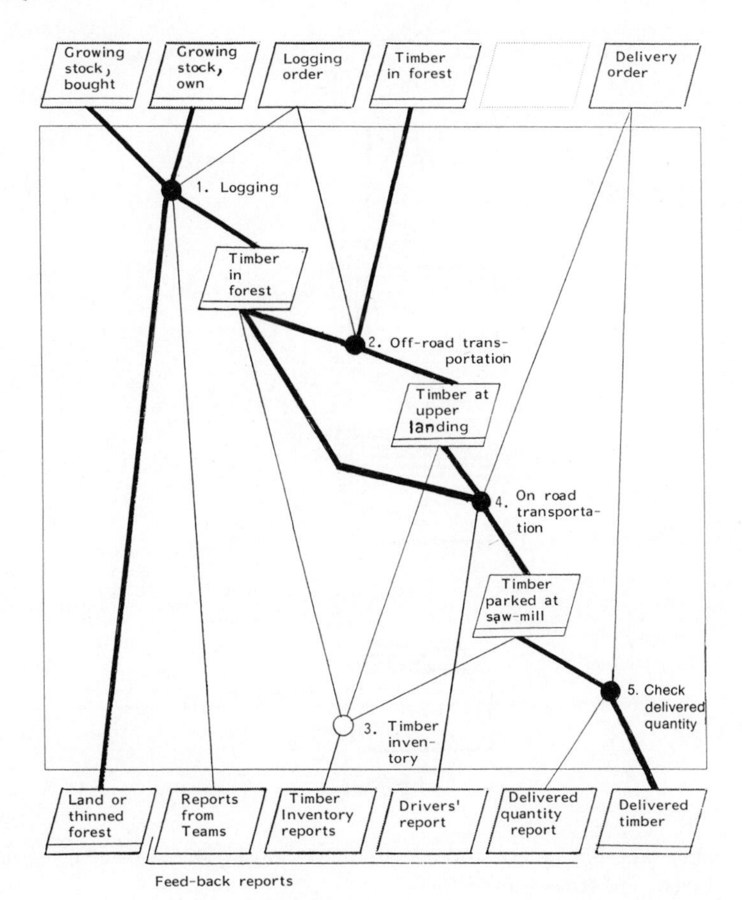


FIGURE 2. The object system for timber production.

questions will vary for different people.

The many problems of dealing with an organisation do not diminish over time, since the view of the people making up the organisation is subjective and keeps changing. We are dealing with a continuously-varying problem situation of unstructured problems in which even the goals are poorly-defined. Is it possible to model such a system? If so, how?

## THE OBJECT SYSTEM—A CONCEPTUAL MODEL

A required assumption for development of an information system for control of forestry production is that we understand the organisation which is responsible for this task—the object system.

We can formulate our understanding in a ROOT DEFINITION and a CONCEPTUAL MODEL (Checkland, 1981). For our object system we can use the following root definition:

A system that produces 'delivered timber' as output using 'growing stock' as

THE VERTICAL ORGANISATIONAL STRUCTURE
OF A ”REGION”.

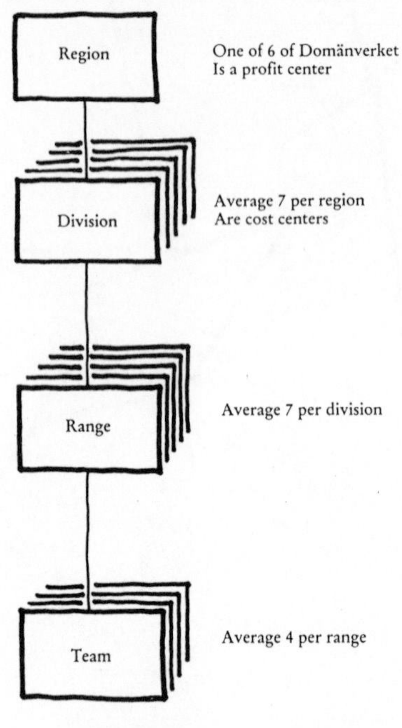


FIGURE 3.

input. The key people in this system are range officers, members of teams, drivers and lumber inspectors.

A simplified conceptual model is given in Figure 2.

The output is not only delivered timber, but also thinned stands, clear-cut areas and four kinds of reports forming feed-back to the information system. Input is also control data from the information system that form the basis for decisions and activities including logging and delivery orders.

The object system has five subsystems (Figure 2):

Logging;
Off-road transportation;
On-road transportation;
Inventory;
Checks on delivered quantities.

#### THE PROBLEM SITUATION

Domänverket Sweden is a profit-making organisation owned by the State. The organisation is divided into 6 regions, each of which is a profit centre and responsible for its own development. The vertical structure of a region is given in Figure 3. Our commission was to develop the information system for operative control of the object system given in Figure 2 and to install a computerised decision support system.

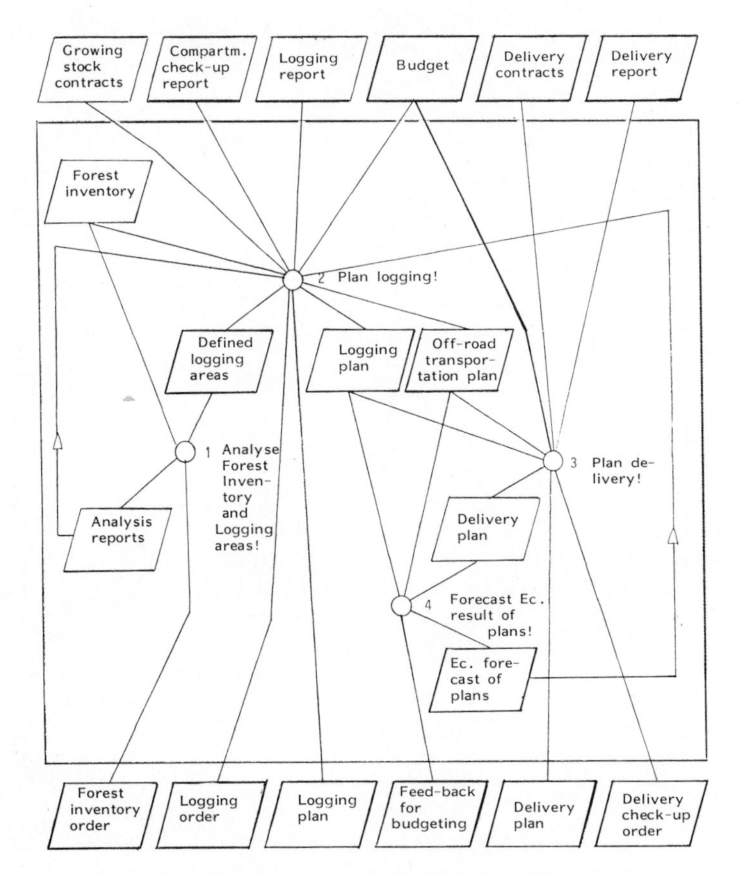


FIGURE 4. Operational planning of timber production.

The main reason for a commission like that is a 'problem situation'. A 'problem situation' is a nexus of real-world events and ideas which at least one person perceives as problematic: for him/her other possibilities concerning the situation are worth investigating (Checkland, 1981, p316). That definition agrees entirely with our previous discussion. The problem situation is subjective and varies both with individuals and over time.

In 1979 our information system was mainly manual and computer assistance was limited to follow-up reporting. These data systems were based on an out-dated basic view which gave too much importance to control exercised by the higher echelons of the organisation, and, in addition, they were batch processing systems. The main problems according to many people were that:

The forest inventory was not accurate enough.

The environment of the object system was so unstable that rigid planning was meaningless.

The capacity of machines and people in the object system was not sufficiently known.

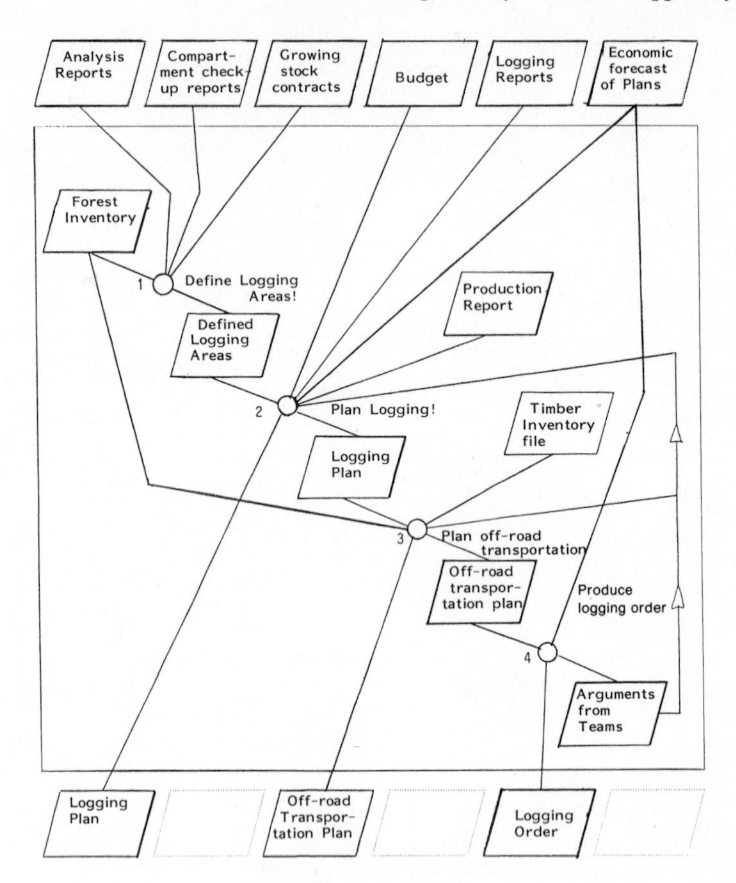


FIGURE 5. Plan logging.

Planning involves many calculations, and it was impossible to make up new plans as often as desired, let alone alternative plans.

The climate (working atmosphere) of an organisation results from perceptions about the relationship between the hierarchy of that organisation and its decision-making procedures. If the climate is not right, then the people who work for the organisation feel that the conditions under which they are expected to work are not appropriate for their jobs. They might consider that they are expected to make decisions in areas in which they have no responsibility, and without being given access to the information necessary for making the appropriate decisions, even though the information may exist. We paid a great deal of attention to these views during the process of developing an improved information system.

## A CONCEPTUAL MODEL OF AN INFORMATION SYSTEM FOR FORESTRY PRODUCTION CONTROL

A workable root definition of our information system is:

An information system, owned by Domänverket Sweden (Umeå Region),

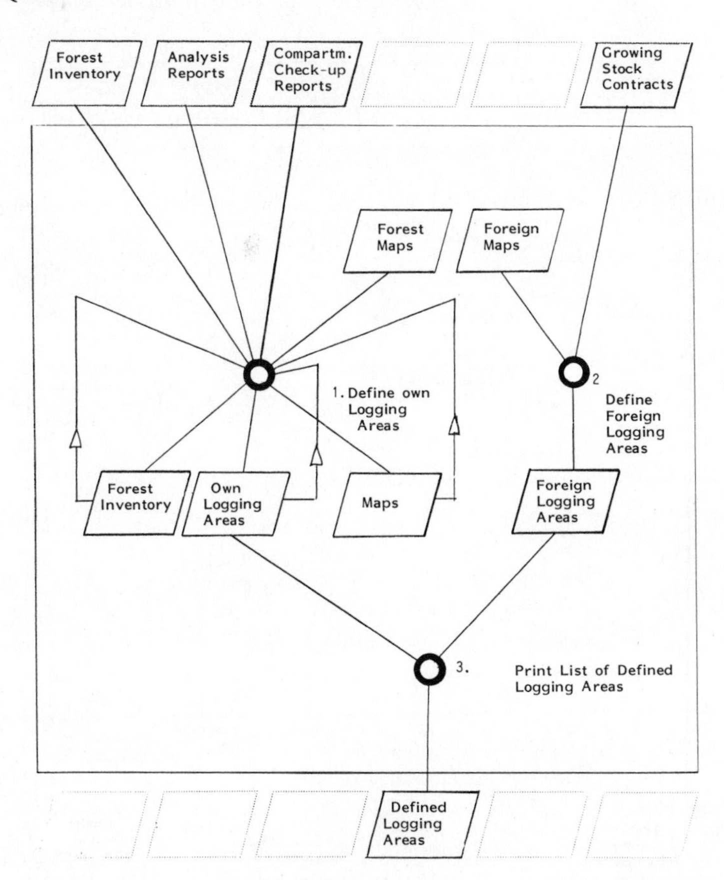


FIGURE 6. Define logging areas.

whose Managing Director has responsibility for development of the system and thus has to decide upon the objectives. The system should, within the constraints set up by the Forestry Protection Law, the policy of Domänverket and other agreements, enable the overall goals of the company to be translated into direct control of the object system for timber production (Figure 2). This transformation is carried out mainly by division officers and range officers and directly affects the teams of the object system, the customers and the management of the region. The basic premise on which the system is based is that positive support for people based on a dialogue about goals is more effective than one which only emphasises follow-up and control of their results.

The conceptual model of the information system based on this root definition is shown in Figures 4 to 7. The nodes in these figures (represented by bold-faced circles) are computer programs, which are all used in an interactive manner. Figure 4 is a brief model of the 'total' system for planning and Figure 7 is a brief model of the 'total' system for feed-back reporting at the divisional level. Figures 5 and 6 are models of

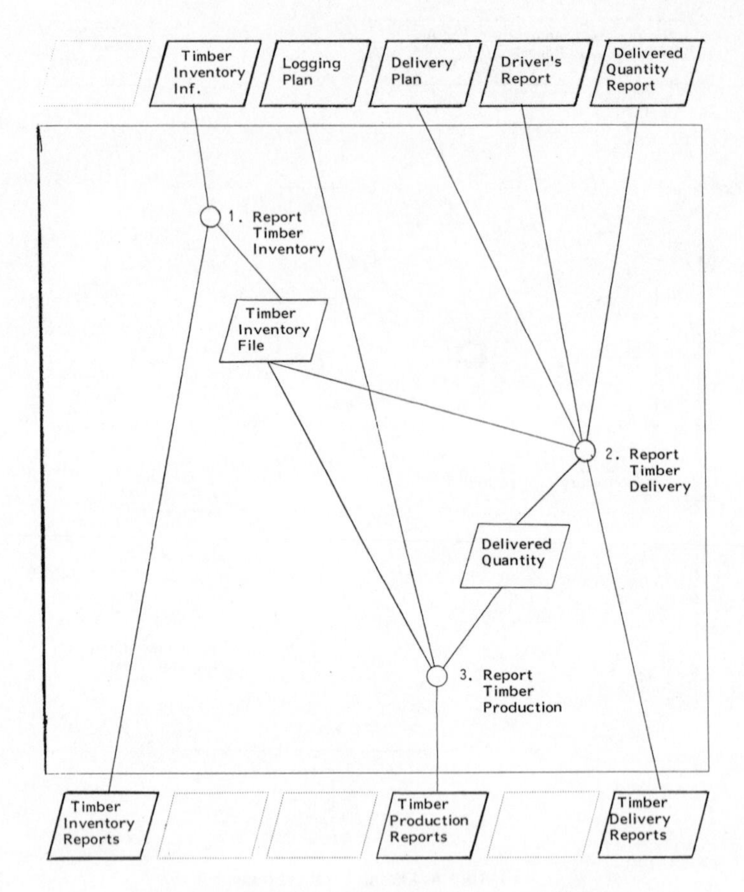


FIGURE 7. Monthly feed-back reporting at division.

sub-systems of Figure 4 on two higher levels of resolution.

DATA SYSTEM CONSTRUCTION

The conceptual model includes a high number of relatively complex computer programs.

When we started the construction of the data system, computer programs and a data base (1980), we tried to simplify the programming and increase the flexibility as much as possible. Therefore we decided to use the best tools available at that time: data base languages, screen painting languages and query languages, as a complement to a traditional programming language (Pascal).

Although we did not succeed in our attempt to decrease the time-consuming programming, the introduction of a query language put us on a track which should prove to be very useful. The query language proved to be a tool that stimulated people successfully to develop their own programs.

## A NEW ENVIRONMENT FOR PROGRAM DEVELOPMENT

New tools have, since 1980, drastically changed the environment for computer program development. They are often referred to as 'fourth generation languages'. Martin (1982, p28) states that a language should not be called 'fourth generation' unless its users obtain results in one-tenth of the time needed to do the same job in COBOL.

Development of an information system of the kind described above is a never-ending learning process. The environment of the organisation changes, people develop and change their views. This environment requires 'user-driven computing', in contrast with 'pre-specified computing' such as is used in, for example, air traffic control and airline reservations (Martin, 1982, pp54-56).

We completely agree with Martin (1982, p56):

> Pre-specified computing needs the fourth generation language wherever possible, because the old methods are too slow and expensive.
>
> User-driven computing needs the new methods because the old methods simply do not work.

This statement is easy to understand from the perspective we discussed earlier. It is not possible, using conventional programming languages and at reasonable cost, to develop computer programs as people change their views.

## DISCUSSION

In our project we spent much time discussing problems with people in the organisation, although we called it 'Specification of User Requirements' at that time. In retrospect we now ask: Did we spend too much time on this? I do not think so. During that dialogue (which did not only concern problems), people's ability to interpret common references, their knowledge about the object system as a system, and their personal views were developed. We developed a common personal view which should prove to be very important in our discussion about feasible computer systems. It is also important now as we install the data system.

One of the first things we found was that an earlier attempt to install a batch processing system for production planning had failed. The reason was that the data processing, including data transportation, took too long to meet the users' desires to change their plans. That system 'died'.

In the present situation we have an interactive system which does not need to die for that reason. But we have today an information system which involves people to a greater extent. We have 'user-driven computing'. That means, hopefully, that people develop faster. This introduces the obvious risk that our data system will die if we cannot develop new computer programs in pace with people's requirements.

To date, it has not been obvious due to technical problems involving telecommunication and communications between different types of computers. People who are busy with technical problems have no time to deal with new ideas about how to plan. However, when the technical problems are solved, and perhaps also a more effective training of the users can take place, the demand for changes will increase. If this cannot be met by our conventional programming tools, there will develop an 'unbridgeable gap' between the model, the computer programs, and people's attitudes. The system will die.

REFERENCES

Checkland, P. (1981) System thinking, system practice. JOHN WILEY AND SONS, Chichester.
Martin, J. (1982) System development without programmers. PRENTICE HALL.

# INTRODUCING COMPUTING TO AN ORGANISATION

G.P. TOTTLE, MA, FBCS

*Agricultural Computer Systems International (ACSIL)*
*Toft Green, Congleton, Cheshire CW12 3QF, U.K.*

SUMMARY
The paper covers the major stages of computer introduction, from initial study to operational implementation. It concentrates on the system, applications software and external user aspects under the following headings: Investigations and Pilot Study; System Proposal/Requirements Statements; Application Development; Programme Management; Field Implementation. Theoretical aspects are illustrated by reference to the author's practical experience in developing agricultural software packages for estimates, smallholder authorities, co-operatives and credit agencies.

INTRODUCTION
This paper is based on personal experience of computer applications in areas which seem to be both relevant and familiar to foresters. These are applications in smallholder development authorities such as the Rubber Industry Smallholders Development Authority (RISDA) in Malaysia, or the Kenya Tea Development Authority.

Due to space restrictions, the scope of the paper will be limited and I propose to concentrate on four factors:

a) Initial Investigation/Pilot Studies
b) Applications Software
c) Programme Management
d) Field Implementation

while omitting other important topics such as hardware selection and installation planning. An expanded list of these and related activities, oriented towards the production of applications software, is given in Figure 1. The remainder of this paper concentrates on the above four factors.

A key point, for those new to computing, is the importance of ensuring these various components of your application are consistent. For example, to introduce a data entry program which has minor differences from the main stream programs which process the data is potentially more damaging than equivalent failures in manual systems, because the usual human safeguard, asking sensible questions, is absent from the chain. More important and less obvious is the need for the consistency at higher levels in the system, and across time. Otherwise it is too easy, two years into the implementation, for the rationale behind the original concepts to be forgotten. In SCAPA (System for Computer-aided Agricultural Planning and Action), therefore, all the documents in this set include cross-references back up the chain to the original statements of requirements. Any changes therefore are proposed and justified in the context of the system as a whole.

Software:
- Investigation
- Pilot study
- Requirement statements
- Specification
- Structure chart
- Data dictionary
- Source code
- DDE (data entry) format programs
- Job control macros
- Helping facilities
- Message library
- Security/recovery aids
- Change control
    - Field Implementation:
        - Installation test
        - System test
        - Beta tests
        - Performance benchmark
    - User Documentation:
        - Management guide
        - Field guide
        - Operations guide
        - Sample forms
        - Sample outputs

Training Aids:
- Video
- Course materials
- Slides
- Trial data files for initial use
- Specialsed courses for
    - management
    - operations
    - field staff

Support:
- Local site support visits
- Customer enquiry
- System maintenance

FIGURE 1. Application system content

I will cover the subject generally by references to examples arising from our implementation of SCAPA with RISDA in Malaysia. SCAPA results from a project set up by ICL's Research Division and Manchester University involving people from many disciplines to examine what contribution computing could make to agricultural and rural development. Their proposals have benefited greatly from inputs and criticism from ODA and the World Bank, from development institutions, consultancies and universities in the UK and overseas, and in particular from visits and detailed studies with development authorities in several countries. A summary of SCAPA's major activities is shown at Figure 2.

RISDA has a very wide remit to promote and support the interests of the 'man or woman behind the rubber tree'—280,000 smallholder families with holdings averaging 5.5 acres (2.2 ha.). Thus, in addition to supporting the mainstream technical activities of replanting, new planting, production and processing, the Authority runs a variety of locally controlled projects with farmers to explore

Major SCAPA Activities

Match farm potential to production plans
Record key activities in plans
Monitor & support achievement
Free extensionists for more creative tasks
Identify specific problems
Allocate credit in relation to farmer's achievement
Plan and direct the provision of inputs
Record sales of produce and provide financial accounts
Summarise management progress for management and other authorities

FIGURE 2.

opportunities for new crops, livestock, new markets, rural industries or trade, and tries to pull some of the industrialised activities back from the towns to the villages. (Leakey and Tottle 1982) This involves an imaginative, project-oriented approach, with RISDA's extension officers (advisers) acting as catalysts in a loosely-structured system—marketing, for example, is organised by RISDA in fortnightly auctions of unsmoked rubber sheet, transacted directly between growers and buyers, with RISDA staff stepping in to bid and purchase only if the price offered seems unreasonably low. By contrast, a similar study in Kenya revealed a much more integrated system (see Figure 3).

INVESTIGATION

The first stage involved the preparation of studies to identify major opportunities and problems arising from the introduction of computers. These studies required holding discussions with all interested parties. The inclusion of opportunities as well as problems is important. In SCAPA we found the academics had generally a highly developed critical faculty, very effective in spotting an organisation's problem areas and the applicability of computing techniques, but very slow to spot areas of opportunity.

To illustrate an opportunity area, as opposed to a problem area: smallholder schemes commonly apply rigid straitjackets on farmers—agronomic packages involving one or at most two seed varieties, standard planting dates, block tractor ploughing, standard inputs and so on. Computing, however, is a management tool which accommodates diversity, and it introduces the opportunity to manage or support efficiently a much wider range of activities—several variants of agronomic packages, for example, or differential rates of credit to favour certain crops and discourage others. In forestry, for example, one could envisage plans and action lists which relate in greater depth and precision to the particulars of topography, soil type, access characteristics and possible tree crops for smaller units than is commonly practised.

These studies resulted in working papers on the various human organisational activities involved, for example:

a) Farm profile (ie. capability) information capture and analysis;
b) functions of extension staff;
c) ordering, scheduling and billing of support services;
d) crop marketing.

Each paper summarised the current procedures, identified potential areas for

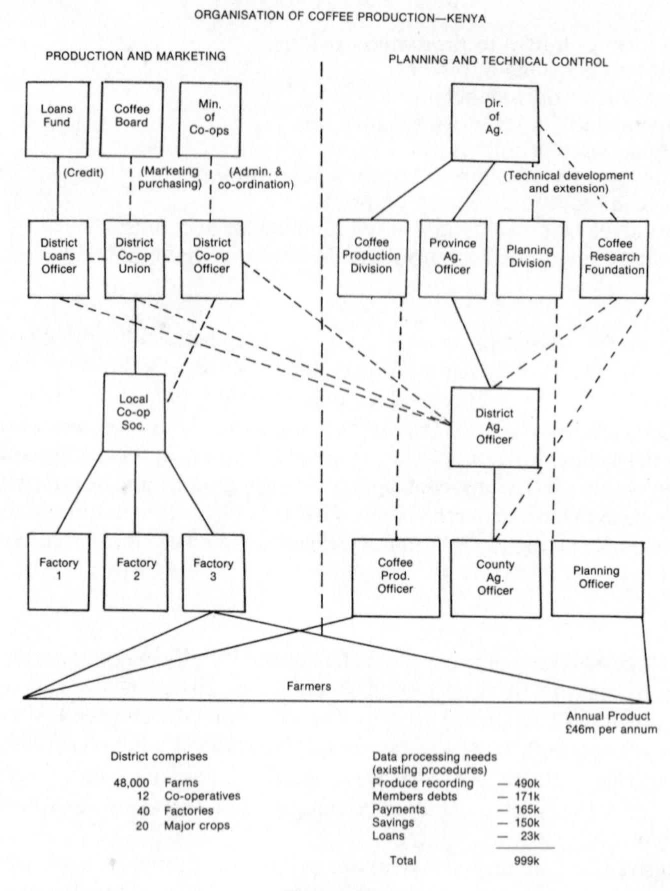


FIGURE 3.

introducing computers, and led towards an analysis of costs and benefits. The Profile paper, for example, included timings and costs for plot measurement, sketch maps, soil profiling and so on. As with other documents the papers were circulated for agreement under formal change control.

PILOT STUDY

It is often appropriate early on to carry out detailed studies from the grass-roots upwards after the basic opportunity and problem areas have been identified in theory. These we carried out in SCAPA in five-week studies in Kenya and Malaysia. The study in Malaysia concentrated on the state of Kedah, a designated poverty area.

During the pilot studies we visited four to eight farms a day and spent one or two hours with farmers and extension and supplies staff, tracking back issues as they arose such as high mortality in seedlings. This could be caused by the late arrival of planting material, or fluctuating demand at nurseries, or a lack of planning information about smallholder intentions. This study process is valuable both for its

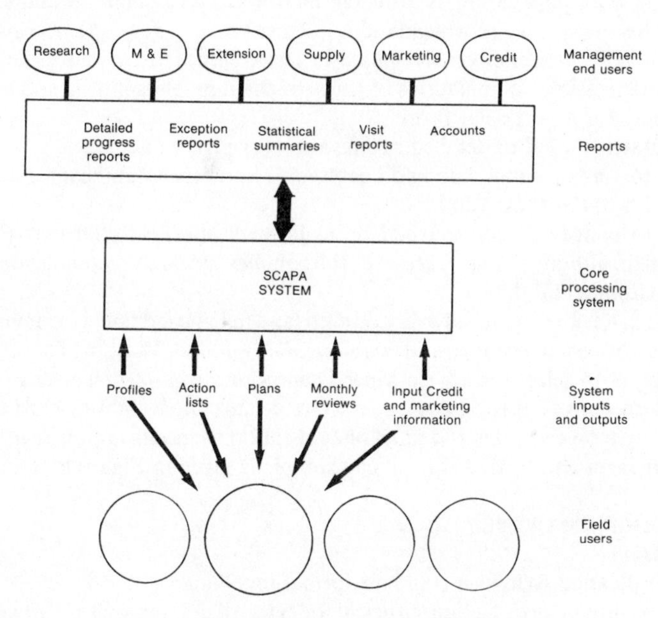


FIGURE 4.

own sake and as a means of obtaining interest and commitment, both among the users and among the implementation team. For us it also helped to understand the information flow and needs, organisational structure and responsibilities. Figure 3 gives as an example the organisational structure in Kenya and Figure 4 the information flow we proposed to map onto it.

STATEMENT OF REQUIREMENTS

Such statements are broadly the distillation of the conclusions from previous stages and their expression in terms sufficient to lead to the computing system and applications software, bought-in or in-house, to meet the resulting requirements. For example, in a similar study in Kenya the following statement on the effective transfer of research recommendations to the field was included:

> *Implementation of Recommendations*
>
> Some of the recommendaions, for example Technical Circular No. 29, Standard Recommendations for Fertilisers, 1976, provides complex guidance on courses of action relevant to particular environments. The potential benefit from this information in terms of improved productivity are probably not achieved for two reasons.
>
> Firstly, data such as the soil characteristics of each farm and of much larger areas, are unknown—the general assumption on soil characteristics is that all coffee-producing areas of Murang'a are very similar. Thus Circular 29,

providing eleven pages of detailed guidance on fertiliser selection and rates and timings of application, is generally interpreted as a single recommendation.

Secondly, the recomendations themselves are very often complex. For example, the section on phosphate fertilisers alone requires a decision on which of 12 possible combinations of levels of calcium, phosphorus and soil acidity apply for a particular farm.

This statement led to detailed specification for programs:

to verify, manipulate and analyse soil and other significant data (PROFILE ANALYSIS)

to convey simple instructions to farmers and field staff (ACTION LISTS) which, although simple, relate to the complex research recommendations and environmental data.

Research advice is frequently very detailed (and full of conditional recommendations which are not easy to unravel) and therefore often ignored. The aim of SCAPA is to use the computer to select the advice which is appropriate in particular circumstances, and then to convey this advice as an ACTION LIST in straightforward terms, for example, with precise rates and times of application of specific inputs related to each particular farm and field. A simplified example is given in Figure 5.

## APPLICATION SOFTWARE

### *Specification*

Each specification follows a standard format involving:

1. A brief introduction sufficient for external users as well as software people to understand the function concerned
2. Input definitions together with a specification of control parameters, a formal description of each input file, record and field.
3. Processing—a description of the function carried out on the input
4. Output—a detailed description of the formats and contents of all output files.

These were followed by other sections covering: exceptional conditions; storage, performance and other implementation factors; potential enchancements.

### *Structure Chart*

Each program was built from a structure chart following the Michael Jackson technique for structured design (Jackson, 1976 and later). I would particularly commend this, both for speed and reliability of implementation since we believe we gained savings of up to 50 per cent. It also gives flexibility in handling modifications with minimal interference with the integrity of the system. (Triance and Yow, 1978).

### *Data Dictionary*

This was used to ensure that all programs worked to the same data structure. For external users it raises particularly important issues about the viability of an integrated database approach to the organisation of their computer applications. In the Rubber Industry Smallholders Authority this integrated approach proved feasible; but in the Kenyan study (see Figure 3) conventional file handling was appropriate, to ensure that each management unit had autonomous control of its own data.

| Period No. | Action No. | Activity | Critical date | Necessary predecessors | Mandays |
|---|---|---|---|---|---|
| 9 | 1 | Prepare seed pieces 30cms long with 3 well developed buds cut from a disease free plantation and deliver to plot (3500 pieces per hectare) | 14/09/00 | | 4 |
| 9 | 2 | Burn trash remaining from previous crop | 15/09/00 | | 1 |
| 9 | 3 | Subsoil with tines only at the centre of each luwang of previous stand of cane using paltik method using 40hp tractor working 2hrs per hectare | 17/09/00 | 2 | 1 |
| 9 | 4 | Plant seed pieces 10 to 11 inches deep and at 45 degree angle. 3 pieces per metre in rows 1 metre apart out closer in wider rows and vice-versa | 22/09/00 | | 0 |
| 9 | 5 | Deliver diuron 85% WP (Karnex) 2kg per hectare | 14/09/00 | | 0 |
| 9 | 7 | Apply preemergence herbicide Karnex at rate of 2kg per ha by sprayer tank-mixed with 2-40 1 gallon per hectare in cane planting row only | 23/09/00 | 5, 6 | 1 |
| 10 | 8 | Mechanically weed and chop stubbles between the new cane rows with a surface chopper (tractor mounted) | 31/10/00 | 7 | 1 |
| 10 | 9 | Deliver 11 bags fertilizer per hectare comprising 3 bags Ammophos and 8 bags Potassium Chloride | 30/10/00 | | 0 |
| 11 | 10 | Apply first fertilizer dressing by cup method to the side of each individual cane plant mixing 2 parts 18:46:0 to 8 parts D:D:60 (3 bags + 8 bags/ha) | 05/11/00 | 9 | 4 |
| 11 | 11 | Second pass with surface chopper between cane rows | 15/11/00 | 8 | 1 |

FIGURE 5. Sugar*Groundnuts Action list

*Data Entry*
I should mention the availability of special packages for this purpose—the effect on system efficiency, ease of use and integrity can be immense; some projects have incurred excess over-runs of up to 25 per cent on their total implementation budget because they neglected this area. (Mackey 1981, Fulton 1981)

*Change Control*
The critical importance of change control throughout the development of an application was highlighted in the first paragraphs of this paper.

*Security Recovery Aids*
Provisions in this area are often of crucial importance, meriting external consultancy. As a check list, the main points to cover are:

1. Definition of file and program access/authorisation levels.
2. Off-line control, including special control of corrections and re-submissions.
3. Retention of transaction files for back-up purposes.
4. DDE standards for input, verification, batch control and possibly locally defined credibility/checks.
5. Regular archiving of main data files, at least weekly.
6. Use of operating system facilities
   —control over users' access to the database, and of their permitted mode of access (e.g. Read-only access to financial information).
   —job control 'macros', which automate the running of sequences of programs, the files they access and the parameters they obey.
   —passwords, which are required by the system, before the user can access privileged facilities for example.
7. Recovery procedures, including periodic practice recoveries.
8. Locally defined audit trails.

PROGRAMME MANAGEMENT

This is an approach to the management of highly-integrated products which has become widely accepted in the software and hardware industries. It sounds obvious and rather tedious, but its absence in the areas of application we studied was, we often felt, responsible for many of the management problems we encountered. Programme management involves the appointment, reporting direct to the overall project manager, of a senior, vigorous, gregarious programme manager (and a group working under him if appropriate) with long experience in the development disciplines.

His (or his group's) functions are to ensure that each unit participating in the programme has clear written definitions of its plans and targets and agreed completion dates, each with stated resource requirements. All these have to be verifiable by, for example, a previously defined and agreed acceptance test and demonstration for each program, against an externally produced dataset. All these are fully documented, and the tests (signed off by the manager concerned in person) gradually integrate into the component tests and system tests mentioned in Figure 1.

Programme management succeeds only if it is regular and rigorously backed up.

Progress in ICL, for example, is reviewed weekly at all levels, taking probably one day of every manager's time. All deficiences or delays are reported upwards, reaching higher levels as their seriousness increases; actions are taken to retrieve or plan round problems, and these also are monitored weekly.

There are tools available to help this, generally based on Critical Path Analysis and Programme Evaluation and Review Technique (PERT). SCAPA Action Lists in fact are one technique particularly applicable to agriculture and forestry—we developed the programs to take account of the fact that the real world is not as precise and predictable as PERT, for example, assumes. The SCAPA action lists therefore are open ended, as, for example, the cane action list at Figure 5.

## FIELD IMPLEMENTATION

The checklist which I would recommend here is that given in the lower half of Figure 1. Perhaps the key point is to have a carefully phased approach, which contains contingency activities which can be re-scheduled or scrapped if needed, but maintains considerable pressure. The pressure is effective for various reasons: to capitalise on training while it is fresh in the minds of the trained, for example; or to minimise the take-on period during which parallel running of computer and manual systems is necessary.

## CONCLUSION

The key note on which I would like to conclude is that the introduction of computing should be seen as an opportunity to re-think established procedures and to seek opportunities which the technology offers. Most organisations start by introducing computer-based versions of established clerical functions, typically payroll processing. This is perhaps a good teething project, but only as long as the more dramatic opportunities for introducing greater diversity in the range of activities which an organisation covers are taken up.

## REFERENCES

1. Fulton, M. (Sept. 1981), A review of experience with nicros in evaluation and project planning. *Agricultural Administration Network Papers,* **13.**ODI., 8.
2. Jackson, M.A. (1976) Principles of Program Design, APIC. Academic Press.
3. Leakey, C.L.A. (1981) The use of computer systems to aid production, management and administration, *Proceedings of the Fourth Conference of the Association for the Advancement of Agricultural Sciences in Africa, Cairo.*
4. Leakey, C.L.A. and Tottle, G.. (1982) A system for computer-aided agricultural planning. Computing for National Development. BCS. 10. Heyden & Son Ltd.
5. Mackey, E.C. (Sept. 1981) Computing in programme planning and evaluation. *Agricultural Administration Network Papers,* **14.** ODI, 19.
6. Triance, J.M. and Yow, J.F.S. (1978) Experience with schematic logic pre-processor, UMIST Manchester, internal paper.

# TRAINING REQUIREMENTS

H.L. WRIGHT

*Commonwealth Forestry Institute, Oxford*

SUMMARY

The introduction of computers into an organisation creates a need for training. This paper looks at some important factors that should be considered in assessing the training requirements. It is stressed that the need for training must be recognised in the planning and design stage and borne in mind when choosing hardware and software. Training methods are discussed both with reference to business software and to more specific forestry applications.

INTRODUCTION

Training plays a vital role in the introduction of computers into forest management and administration. Together with the correct choice of hardware and software it will contribute to the efficient running of the system. The aim of this paper is briefly to outline some of the important factors that must be considered when planning the necessary training since too often this aspect is neglected or relegated to a low priority. This is unfortunately reinforced by many books currently available with such generic titles as "Introducing Computers to the Small Business" or "Management Using Computers", which pay little attention to training although giving great emphasis to the choice of hardware and software. The documentation produced by the suppliers of the latter cannot usually be used for training as a glance at the shelves in the computing section of a bookshop will reveal. Here you will find literally hundreds of titles, many of them aiming to interpret for the non-specialist the manuals produced by the manufacturers. Software packages exist which supposedly make it easier to use other software packages, surely an indication of failure on someone's part.

The training requirements can be considered under several main headings. The need for training must be recognised before any computer system is designed, chosen and installed. The type of training will also depend on the specific use of the computer within the organisation. For forestry the uses will be those associated with running a business and those that are specific to the practice of forestry and its allied industries.

CHOICE OF HARDWARE AND SOFTWARE

It is not the intention here to discuss the choice of hardware and software except where it influences the degree and intensity of training required. There has been much written, often in advertising material, about 'user-friendly' operating systems and program packages. To quote a recent example for an integrated business package 'It isn't just user-friendly, it's positively sociable!' Many computer systems and applications packages have been, and, in some cases, still are distinctly unfriendly to the user. The communication between the user and the machine may involve complex operating commands and codes, and technical jargon incomprehensible to the non-specialist. Peltu (1984) gives a set of useful guidelines for judging the 'user-friendliness' of a system.

The amount of training required can be very much reduced by a well-designed system. Managers should read reviews of potential software, talk to other users and try to obtain actual experience before choosing a system. The description in the advertisement '******** is easier to learn and simpler to operate than programs that cost half as much' may be offset by a reviewer writing 'Despite the online tutorial and the apparently friendly appearance of the manuals it is really not that easy to get to grips with ********'. It is important that users rapidly develop confidence in their ability to communicate with the system. A bad initial experience can induce an antagonism which may be long-lasting and difficult to remove.

Potential users at all levels should be consulted and, if possible, exposed to various systems during the planning phase before a final decision is made. Armytage (1984) quotes the manager of an Integrated Electronic Office Trial Programme as saying 'It is easier for the office worker to understand technology than it is for an analyst to understand what goes on in the office'.

### WHO NEEDS TRAINING?

The first picture that comes to mind is of computer operators sitting in front of video terminals all day entering data or of a typist using a word processor. However, increasingly middle and senior management are finding that direct access to computers via their own keyboards is giving them needed information. For example, the use of a financial modelling package, combined with graphical output, assists them in making decisions. Obviously the main thrust of any training programme must be geared to those concerned with the actual operation of the system. But there are still many misconceptions about computers and these may need to be dispelled amongst the workforce as a whole. Megarry (1984) points out that some employers confine their training and briefing to a small portion of the workforce, thus failing to realise that the introduction of a computer is of far wider significance than simply to those who will work with it.

### BASIC REQUIREMENTS

Whatever machine is installed someone will need to be familiar with the operating system, that is, the vital program that is in overall control of the computer whenever it is operating. Without it the computer could not load programs, process data, retrieve or store on disc or send output to the printer. Documents or information have to be filed and and retrieved, created, updated or destroyed, copied or printed in an automated system just as in any manual office system. Converting a chaotic manual system usually results in automated chaos since the computer is no guarantee of an efficient system.

It is essential to develop an operating discipline and to instil this into those who will use the system. This training will include:

- basic instruction in starting and running the system, such as what to do in the event of power or other failure;
- operation of peripherals such as disc drives and printers;
- simple maintenance;

procedures and precautions in disc handling, the protection of disc file security, and so on;
maintenance of regular back-up copies;
methodical filing of discs, labelling and written catalogues of each disc;
general disc housekeeping with regular deletion of unwanted files, and secure storage of master copies.

Ideally this should be documented and available to all users. It can then serve as training material, something for which the manufacturer's manuals are seldom designed.

If a large system is being installed the suppliers may include a certain amount of training in the overall contract price. Access to expert advice should always be available and some suppliers of applications software will run a 'hot-line' service for instant trouble-shooting.

### USE OF BUSINESS SOFTWARE

The use of generalised packages for specific business applications forms a major part of many computer systems. The commonest type of Package are Accounting, Financial Planning (Spreadsheets), Data Base Management Systems and Word Processing. The latest trend is to integrate these individual applications, particularly the last three, into a single package. Whatever packages are used, it makes sense to try to ensure that they are as compatible as possible with the same control commands and the ability to switch files from one to the other. This has an obvious benefit in training.

It is important that those using the packages are aware of the function of a particular package. This will be to a greater or lesser extent, dependent on the responsibility of the person concerned. Robotic use of packages does not make for efficency.

### FORESTRY APPLICATIONS

Concurrent with the development of more integrated packages has been an increase in the software commercially available for specific groups of users. Few exist for specific forestry applications. An increasing number of programs are being published, particularly in the North American literature, but many of these are for very specific local conditions. However, they may form the basis for conversion to other conditions. Most software thus needs to be written for a particular application although business packages can be used for certain tasks. For example, a data base package could maintain a record of compartment histories and produce regular tables of areas, volumes by age-class, species and so-on. Some spreadsheet packages have built-in functions to calculate net present value which make them attractive for financial planning in forestry.

The need to develop software may require a greater programming expertise within a forestry organisation as commissioning specialist software for a specific application from outside agencies can be very costly. The implementation of home produced software may involve considerably more time spent on training than that required for simply using a package.

TRAINING METHODS

Training can range from a formal course provided by an outside organisation to some informal help from another employee. Regardless of the form of the training the same basic principles apply. These are given by Megarry (1984) as:

Practice precedes theory: let trainees gain experience of the equipment before giving theoretical explanations of what will happen;

Theory precedes practical: once trainees have had 'hands-on' experience, make sure that they understand what was happening and why;

Combine the above two principles to provide an alternative between practice and theory, so that practical experience makes the general principles come to life and knowledge of the underlying principles prevents practice from becoming superficial or mechanical;

Never try to do too much at one time. Until trainees have confident mastery of one level, they will be unable to digest new ideas at a higher level. The kind of questions people ask is a good guide to the level they are operating at; if they are not asking questions, make few assumptions about how much they understand.

Trainees should be able to work at their own pace and not be forced to keep up, or be held back, by slower learners. This requires suitable training documentation which allows individuals to choose their own fate and to be able to work on their own if necessary. The manuals supplied with the system or package are generally useless for this purpose. A set of reference cards providing a check-list of the commands needed is useful for operator training.

Some packages will come with their own tutorial or Computer Assisted Learning (CAL) software that instructs the new user. This will often involve a simulated application complete with sample files thus allowing the trainee to work without the fear of making disastrous mistakes in a real situation. Even if not provided, a set of such sample files can be created for introductory training. CAL software can provide feedback to the trainer to help highlight particular problems. Most software packages now provide HELP facilities enabling a user to ask the computer for information on commands and their format without interrupting the particular job.

All users should have a general introduction to the computer and its operating system, disc housekeeping, before training in the particular application they will be using. Facilities should also be available for extra training once the initial period is over, possibly a half-day per month. The amount of training will vary for different applications, experience and aptitudes. For word-processing it may be a two-day introduction, followed by a week of training and then a regular half-day session once a month. Trainees should not be expected to work entirely on their own even with the best computer-based tutorial system. Someone in the organisation must be made responsible for training. An experienced user is often far better than a computer professional.

Training in programming takes more time. Although John Jeffers used to say that he could teach anyone to program in two days, it takes considerably longer before any serious software can be written. Our experience in Oxford is that it takes from three to six months to train someone with no experience to the stage that he can produce relatively complex applications programs.

Many commercial courses are advertised in computer magazines covering a wide range of subjects. They are aimed at all levels of expertise and experience. In choosing a particular course, care must be taken to ensure that it is at the right level and covers the subject required. For short courses of one to five days, which applies to most of those advertised, costs are from one to two hundred pounds per day. Some training companies will prepare a special course at no extra charge and they can be held on the customer's premises. For a small organisation an investment in such training will need to be carefully evaluated. It could be that it is the best way of ensuring that the person responsible for training within the organisation obtains the necessary experience. Local Technical Colleges and Colleges of Further Education now offer day-release and evening classes in computer science, programming and to a lesser extent applications software.

REFERENCES

Armytage, J. (1984) From silicon shock to the glimmer of a Grand Design. *The Guardian,* November 15

Megarry, J. (1984) *Computers Mean Business.* Pan Breakthrough Books, London. 336p

Peltu, M. (1984) *The Electronic Office.* Aeriel Books, BBC, London. 180p

# PART II

# CURRENT USES OF COMPUTERS IN FOREST PLANNING

# FOREST PLANNING AND CONTROL— A BRIEF REVIEW OF RECENT DEVELOPMENTS

R.T. BRADLEY

*Director, Harvesting and Marketing Division, Forestry Commission, Edinburgh*

## INTRODUCTION

The four papers immediately following this one deal with the planning and control system as applied in different types of organisation. The four papers thereafter deal in greater detail with component elements of the system such as the creation and maintenance of a data base, the development of growth models and the extension of these to economic models covering a wider spectrum. As an introduction to these papers, I wish to review briefly the developments in the planning and control systems used in the Forestry Commission over the last 25 years. Our experience is probably typical of that in other large forestry organisations in this and other countries.

## DEVELOPMENTS IN PLANNING AND CONTROL SYSTEMS IN THE FORESTRY COMMISSION

### *Planning systems*

1960. Almost all planning and control was exercised through individual FOREST WORKING PLANS (that is, planning 'from below').

1965. CONSERVANCY PLANS (Regional Plans) were introduced in an attempt to draw together and coordinate the planning process (that is, an 'intermediate phase').

1970. A CORPORATE PLAN (for the organisation as a whole) was developed to bring broad planning and target-setting to the right level in the organisation, namely at headquarters).

1975. A POLICY ANALYSIS AND REVIEW system was developed to enable Regional (Conservancy) modifications and interpretations of national policy to be discussed formally.

1980. Greater emphasis was placed on the now computerised FINANCIAL CONTROL SYSTEM (partly to get round cumbersome modification of formal plans).

Along with the above developments in the formal planning processes there were parallel developments in the budgeting, data processing and control systems.

### *Control systems*

1960. A MANUAL DATA BASE was used for both growth prediction and financial forecasting.

1965. COMPUTERISED YIELD FORECASTING was introduced to relieve staff of the very onerous work of yield prediction. The Growing Stock data base was computerised.

1970. A MANUAL FINANCIAL CONTROL system was introduced for the organisation as a whole which supplies both management control information as well as strictly financial data.

1975. FINANCIAL CONTROL was computerised on a batch-fed, agency-controlled central processor with manual (postal) submission of basic management and financial data.

1980. Increasing difficulties were encountered with slow overall response time, with postal delays aggravating inherent difficulties involved in batch-processing on agency machines.

1985. A DISTRIBUTED MICRO-PROCESSOR system was introduced to give first-level supervisors computing power to process information relevant to that level and to feed into the new in-house central processor which will deal more with the strictly financial side of the system.

*Discussion of developments*

There is an apparent and interesting contradiction between the increased centralisation of the planning process and the decentralisation of the data processing and control process. This partly reflects the increasing availability of cheap micro-computers with dramatically improved capability; but these diverging trends would be sensible even if the hardware improvements had not been so dramatic. This is because planning processes should in any case start at the top of the organisation and be disseminated downwards and outwards with appropriate modification and interpretation. Data collection at the other extreme has to start at the lowest levels and extremities of the organisation. It was an oddity that the availability of relatively cost-effective central processing for a time distorted the more logical approach of progressive analysis and reducing quantities of data at successively higher levels in the organisation. Other technological developments in the telecommunication of computerised information have, of course, made it more practicable to revert to the more natural distributed processing systems now being introduced into the Forestry Commission and many other forestry organisations.

The Forestry Commission's distribution system is described in more detail in Mr Pritchard's paper. Mr Wheeldon is reviewing the computing requirements of a large forestry management company and Dr Williamson those of a smaller company or estate.

Mr Westerling's paper describes a Swedish system designed to be available for purchase by individual companies or management organisations. This paper is particularly valuable in that it demonstrates that the problems are much the same in other countries and that the proposed solutions are also along familiar lines.

There is no doubt that the recent and very great technical advances in micro-computing together with very competitive costs have opened up a whole new range of possibilities for the forestry industry. The geographically dispersed nature of most forestry organisations previously imposed great stress on the single central processor approach to computing. Delays in obtaining results, the need for tightly standardised input documents and the lack of a simple interrogative facility meant that the field manager felt that he obtained little benefit from the mass of data fed in and the reams of paper print-out produced long after the management decisions had been taken.

We now have in our grasp the opportunity not only to remedy the defects of the older systems but also the opportunity to make available new techniques to aid the line manager in the field of operational planning and control. The papers which follow describe some of these opportunities and how best to take advantage of them.

# THE USE OF COMPUTERS IN OPERATIONAL PLANNING AND CONTROL IN A LARGE BRITISH FORESTRY COMPANY

H.A.W. WHEELDON, BSc, MIC For.

*Economic Forestry Ltd., Forestry House, Brewery Road, Carmarthen, Dyfed*

SUMMARY

Operational planning within a large company is essentially an exercise in the manipulation of information. Modern computer systems provide the means to achieve these ends, quickly. accurately and cheaply. Systems should be chosen to suit the structure and needs of the company, with the emphasis on simplicity and flexibility. It is considered that the micro-computer is the most versatile tool to meet all the requirements of a large British forestry company. Potential users should plan carefully to identify and meet their own particular requirements.

## INTRODUCTION

Operational planning within a large British forestry company is essentially an exercise in the manipulation of information. The typical large British forestry company can be summarised as follows. It will provide a full range of forestry-related professional and commercial services to private, public, commercial and institutional clients. It will have an operational presence in all geographical areas of Great Britain. It will have a management responsibility for at least 60,000 hectares of forest, and will employ at least 50 qualified forest managers.

Operational planning within such a company will relate to all aspects of its organisation. The purpose of this planning will be to provide adequate staff and resources to each area of activity to permit proper completion of the company's work programmes and obligations. Operational planning will cover forward planning to produce budgets and estimates as well as short term planning to respond to changed circumstances as they occur. If one is to undertake planning and control of the company's operations, whether in the board room or in the forest, the relevant information needs to be readily available and easily assembled and collated.

Traditionally, foresters have relied on written records for management of their forests and their commercial activities. These written records, such as forest compartment records, management reports, valuations, accounts, ledgers, and so on, have provided the basis for forward planning and decision making. Reliance has then been placed on the mental ability and agility of the managers to use the available information to take the correct decisions and achieve commercial success. For the larger company the sheer volume of information, often stored in scattered locations, makes this task harder to achieve with each increase in the level of activity. The senior levels of management, in particular, could become isolated from much information that could be used to assist in the company's success. It is within this context that the use of the computer is not only beneficial but essential for successful and efficient operation.

AIMS

For the large company the aims in using computers for operational planning are broadly to increase both the speed and accuracy of storing, transmitting and assessing information. The information to be handled will differ in context and importance with different levels of management. These levels of management can be best summarised as follows:

a) The Forester or Field Manager who is concerned with the planning and execution of field operations. This includes the organisation and supervision of labour and machinery, the purchasing of materials and the provision of professional services.

b) The Commerical Manager who is concerned with the management of the business as a whole. He controls and supervises the Field Managers and their operations.

c) The Company Director who is concerned with the strategic planning and direction of the business, the control of finances and the development of new and existing lines of business. He will use the Commercial Managers to carry out his plans.

The aims in using a computer to aid operational planning will then be to assist positively all three levels of management to carry out their own tasks, using the same source of basic information.

For the Forester or Field Manager, operational planning is a day-to-day activity, taking into account the longer term objectives or limits that exist for the particular property, contract or operation he is dealing with. He will still expect to rely mainly on his own mind and memory, possibly backed up with a notebook for daily decision taking. He will turn to the computer for accurate recording and storage of the actual events or transactions that have occurred. He will use the computer to update his own memory or notes on a weekly basis, and use the information to project forward likely timetables, budgets or other forecasts based on progress to date. At budgeting time he will turn to the computer both for a record of the previous year's activity and as a means to prepare his own estimates and forecasts for the coming year.

For the Commercial Manager the computer plays a more important part in his short and long term planning. In the short term he will use the information from his individual operating centres to build up a picture of problem areas, strengths and weaknesses. He can then apply his own experience to take advantage of strengths and to correct weaknesses. In his budgeting and forecasting role, he will use locally recorded information to establish a 'picture' of the company activities and provide an overview in terms of estimating and forecasting performance.

For the Company Director the aim of introducing computers is to provide the service outlined above to his Field and Commercial Managers and then present simple consolidations of their own individual plans, forecasts and results into meaningful company records and statistics. He will use the mass of information available to plan areas for expansion, identify growth and problem areas and calculate resource requirements in both cash and manpower.

The use of computers can greatly assist all levels of management by permitting wider and more detailed use to be made of relevant information. All levels of management can draw on the same basic data for different purposes. It can be available very quickly and it can be used concurrently in different locations; for example, a Forester can use the computer to provide details of work in progress, his Commercial Manager can extract similar information to calculate profit margins, whilst the Company Director generates a report on capital expenditure to date. These processes will draw on the same information and are not interdependent. It is this ability to provide a single input—multiple output information service that puts the computer far ahead of any manual system. It is also, I believe, the most telling argument for the use of computers and will probably be the main criterion on which the larger forestry companies will take the decision to proceed with their introduction.

METHODS

Having established that the aims of the company can be met by using computers, how is it done? There are probably as many answers to this question as there are permutations of hardware and software suppliers. The answer for any individual company must lie with its own staff and organisation.

The use of computers is a method for the manipulation of information. To succeed it is important to define very carefully the units with which you wish to work. This is generally achieved by detailed information as a series of records, each record subdivided into a number of facts, such as area, value, and yield class. These individual facts are commonly called fields. Each record will refer to a specific item like a sales transaction. All records will be coded clearly to identify the sector of the company in which they have occurred, and most will be quantified numerically. Given that these records are carefully defined, the other main requirement will be the facility to access and manipulate them at will. Within a company, most information will need to be produced in a predetermined form as reports of one sort or another. The ability to amend reports and produce new ones is essential for the successful use of a computer system. These simple provisions will provide a good starting point from which methods of operation can be developed to reflect the structure and organisation of the business concerned.

It may well be that in designing a computer system to meet the needs of the company, the existing structure and organisation of the company itself may need modification or even drastic change. The common forms of company structure are seen as follows:

a) The Head Office type. In these companies all information is sent to a central office for compilation and computation. This central office then disseminates reports and data as required to all levels of management.

b) The Regional Office type. Other companies are split on a geographical or functional basis with two or more offices providing information services as (a) above. Information is passed down to local offices, and in summary form up to higher levels of management.

c) The Local Office type. With this type of structure, each operating office is responsible for the generation and management of its own information. Normally, only summary information is provided to higher levels of management.

In choosing a suitable system the main requirements are that:

a) it reflects the existing or desired structure of the company

b) it provides information where it is needed (and no more than is needed)

c) it is cost effective

and

d) it works!

To make it work requires simple and logical specification. The four key elements are, I believe:

a) Single point of entry of information from source documents. This will help avoid coding errors, input errors or double entries, and speed up the recording of information.

b) Retention of detailed listings of all transactions, to provide easy auditing, coupled with the ability to access and sort these listings. It is essential that managers at all levels can check the accuracy of information, irrespective of whether it is a pure archival record or whether it is to be used in the preparation of reports, estimates or budgets.

c) Simple reports, using the same basic data with differing degrees of consolidation to meet the requirements of different management levels. Thus a property report may take three forms for different management levels:

i) for the Field Manager a full list of transactions for each property;
ii for the Commercial Manager a summary list of transactions by property;
iii) for the Company Director a summary list of transactions by groups of properties.

It is very easy to demand sophisticated reports, combining and consolidating different variables. Computers can and will produce them, but the end result is likely to be misinformation. Reports should be kept simple and good managers employed to do the extrapolation and interpretation!

d) The last and most important key point is accessibility. The use of computers for operational planning is to facilitate the supply of usable information to managers. It follows that all operational offices will require computer facilities for them to achieve these ends. Modern technology can provide the equipment readily and cheaply. It is important to ensure that all staff have easy access to their requirements.

## THE EQUIPMENT

Consideration of the use of computers requires some knowledge of the equipment and its benefits or drawbacks. I am not an expert but would summarise the situation for the large company as follows:

### *Mainframes*

It is doubtful that any forestry company in Great Britain is large enough to consider the sole use of a mainframe computer for operational planning or any other purpose. Use could be made of a computer bureau, and through it a mainframe computer.

There are substantial drawbacks involved, specifically:

a) *Time*. Normally input will be in batches, probably on a weekly basis with a 7-10-day turn-round time, so that information presented to managers will be at least 14 days old, probably a lot more.

b) *Detail.* Although one can expect to receive audit listings for checking, most figures will be carried forward as 'to date' figures.

c) *Accessibility*. It will be very difficult to gain access unless it is used in conjunction with compatible micro-computers.

d) *Cost.* Notwithstanding (a), (b) and (c), it is likely to cost as much if not more than the other options.

Mainframe computers can be used in concert with micro-computers 'outstationed' at operational offices of the company. These micro-computers can be used to record and transfer data to the main computer, and to receive computed results. They will also provide a medium for local report generation and data manipulation. Such an option can be impressively efficient but also very expensive. The costs of mainframe software, particularly specially produced programs, will be prohibitive. The acid test for this option must be whether the mainframe can offer any advantages to the user over the other options. My view is that there will be no advantages to the forestry company choosing to use a mainframe.

*Mini-Computers*

This is a viable option for a large forestry company in Great Britain.

The use of a 'mini' will suit those with a 'Head Office' type of structure, particularly where centralised accounting is required. Combined with compatible micro-computers in operational centres, all the benefits to be expected by using computers can be provided. In addition, the 'mini' can provide a suitable vehicle for the storage and use of statistical information on a large scale.

One should carefully examine the costs involved, particularly in comparison with Option 3 below. The additional costs in hardware, manpower and particularly in communications should be carefully weighed before choosing a mini-computer as described above. I see no role for a company using solely a mini-computer at a Head Office. One could ascribe to that option the same drawbacks as pertain to a mainframe, with perhaps some reduction in the cost.

*Micro-Computers*

The advent of the micro-computer and its recent rapid development is as important and relevant to the large company as to the small.

It provides a basis for automation of the information gathering, reporting and accounting functions, no matter the size, structure or geographical spread of the company. Modern communications network can provide rapid interchange of data between divisions of a company, or within the divisions themselves.

A careful assessment has to be made as to the requirement in any particular area before a micro-computer is chosen. Some of the smaller and older machines can have considerable drawbacks if required to carry a heavy workload. Currently, however, major manufacturers can offer a varied array of 'modular' designs which permit the capabilities of the machine to be expanded and extended (within quite wide limits) as required. The micro-computer in one of its various guises will probably be the preferred way forward for the large forestry company for the next decade, justified mainly on grounds of flexibility and low cost. As specifications improve, the relevance of the mainframe recedes further and even the mini-computer will have limited application.

THE SELECTED EQUIPMENT

My own company is currently finalising its own plans for the introduction of new computer equipment.

We have for many years been using a mainframe computer through the services of a computer bureau. Latterly information has been relayed to the bureau by telephone line from our own micro-computers. These have also been used in a limited way for other management purposes. After a careful and thorough study we have chosen a system based on powerful micro-computers based in our larger operational offices. The machines involved are modular in design and will vary from small micros with 256K memory and 1MB floppy disc storage, to multi-user micros with 1MB memory and 40MB hard disc storage.

We believe this system will give us maximum flexibility, improved data capture and faster report generation. It will also provide all levels of management with high quality information on an improved timescale. It suits our geographical locations, our functional organisation and our management style.

## CONCLUSIONS

I have no doubt at all that a competent modern computer system is essential for effective operational planning in a large British forestry company. I have tried to cover the benefits to be expected in the preceding sections. To 'prove' the case for introducing computers, one has only to consider the alternative—a 'manual' system and its implications.

It is a daunting prospect to think in terms of manual preparation of accounts, manual record keeping, manual estimate preparation, manual budgeting, manual forecasting, manual consolidation together with all the drawbacks of slow turn-round, difficulty of access, lack of detail, difficulty in checking and the staff requirement.

That in itself convinces me that there is no alternative.

# OPERATIONAL PLANNING AND CONTROL IN THE FORESTRY COMMISSION: THE FOREST DISTRICT COMPUTER SYSTEM

M.A. PRITCHARD

*Forestry Commission, Edinburgh*

SUMMARY

The major operational level of management within the Forestry Commission is at Forest Districts. The responsiblities of this level include planning, control and administration. In 1983 a Working Party examined the data requirements associated with these functions and recommended a micro-computer based system to meet these requirements. Multi-user software was developed for four main components—to monitor wages expenditure (including the calculation of gross to net pay); to monitor the operational performance of vehicles, machinery and equipment; to assist in job cost control; to record timber stocks produced by harvesting operations.

The system will be installed at all Forest Districts from early 1985. Initially the system will run 'stand alone' (ie unconnected to other systems) though communications software is under development with a view to introduction in early 1986. Additional components will be developed in response to demand to meet other operational planning and control requirements.

## INTRODUCTION

The Forestry Commission is organised into three management levels. The current management structure is summarised in Figure 1. The Headquarters in Edinburgh is primarily concerned with policy matters, liaison with other Government Departments and with Ministers and with strategic planning. Certain central service divisions such as data processing and personnel are situated in Headquarters.

The seven Territorial Conservators are broadly equivalent to regional general managers and are primarily concerned with tactical planning and control. There are, however, a number of support functions carried out at the Conservancies such as industrial staff recruitment on behalf of the Forest Districts.

The seventy Forest Districts are each managed by a senior professional forester. The number of Forest Districts range between seven and fourteen per Conservancy. The Forest Districts are primarily responsible for operational planning and control. There is considerable variation in workloads between individual Forest Districts and Table 1 summarises this for a range of functions.

Within most Government Departments there is a body known as the Information Technology Steering Committee (ITSC) which advises on the development and application of information technology in the Department. In the Forestry Commission the ITSC advises the Commissioner for Administration and Finance on policy matters and also oversees systems developed carried out by the Data Processing Division or by individual users. In January 1983 the ITSC formed a small working party to investigate the use of micro-computers at the Forest District level of management.

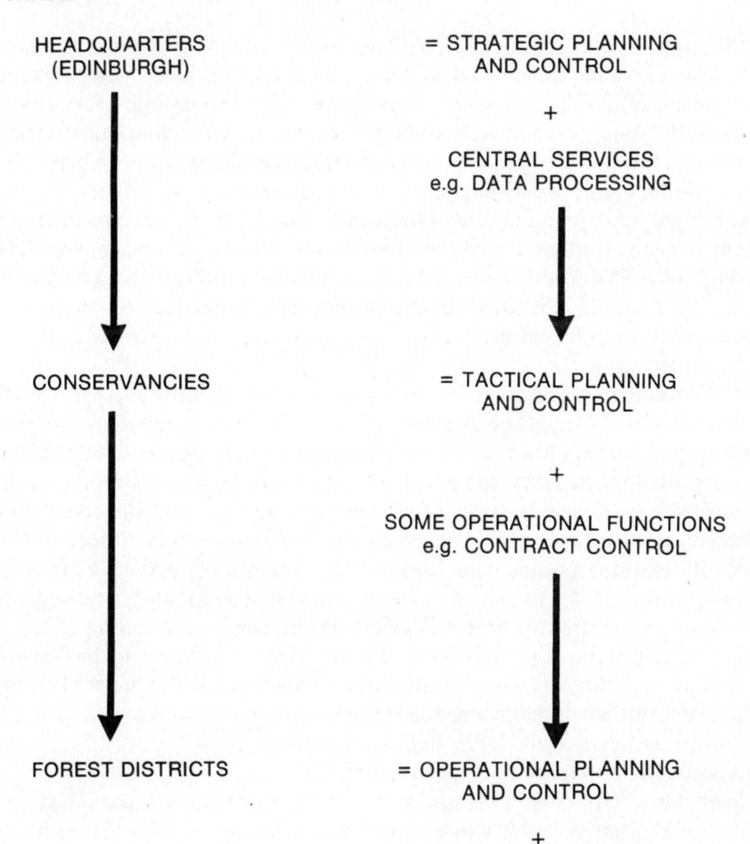


FIGURE 1. Forestry Commission management structure.

TABLE 1. Forest district workloads
(Note that each column should be considered independently)

*Range of values for districts*

| | |
|---|---|
| Plantation area: | 4,000-40,000 ha |
| New planting: | 5-1,000 ha. $yr^{-1}$ |
| Restocking: | 10-520 ha. $yr^{-1}$ |
| Volume harvesting: | 4,000-220,000 ha. $m^2 yr^{-1}$ |
| Professional/Technical Administrative staff: | 7-33 |
| Industrial staff: | 12-160 |

Although all management levels may perform operational functions in addition to their strategic and tactical roles, Forest Districts are the major operational management level in the Forestry Commission. The responsibility for utilising the resources of labour (both directly employed and contract) and machinery rests with the Forest District Manager and his staff. This operational responsibility involves three main functions—planning, control and administration.

At present, data requirements of these three functions are met almost entirely by clerical systems. Certain accounting information is provided on a bureau basis by Headquarters. The Forest Districts do not currently have direct access to computing power. As a result, considerable clerical effort is expended to ensure that the professional and technical staff obtain this information and often the clerks are the foresters themselves.

The Working Party reviewed the three functions and identified the associated data requirements of management. A pilot study using a micro-computer was developed to test the practical application of these requirements. In December 1983 the Working Party reported to the ITSC recommending that micro-computers should be used at Forest Districts to meet their data requirements and to provide key data for other management levels (Forestry Commission 1983). The systems developed were tested in two offices and as a result of this the Forestry Commission has decided to install the systems at all Forest Districts as part of a major distributed processing project. Installation of the seventy Forest Districts will be carried out during 1985.

The purpose of this paper is to review the main factors influencing the design of the system and to outline the components currently developed. The paper also suggests areas where further development will occur.

## DATA REQUIREMENTS OF FOREST DISTRICTS

*Operational Planning.* Essentially this function requires the evaluation of a number of alternative courses of action and a decision based on the results. There may be several different sources of basic data (such as growing stock records, price-size relationships, work study standard times and so on). Frequently evaluations can be carried out well in advance of applying a resulting decision and hence the planning function is not necessarily time-critical. Operational planning is a good example of 'what if?' applications where the interactive use of a computer is valuable.

*Operational Control.* The major feature of this function is that data relating to the performance of a job are compared to some pre-determined standard (usually derived as a result of the planning function). Measures of cost (total and unit) and output are the more common factors that are used. Operational control is usually time-critical in that balanced evaluations and decisions need to be made quickly. The cost of collecting basic data is frequently a constraining factor.

*Administration.* At Forest District level the major administrative function is associated with the running of payrolls and provision of accounting information. Many of the source data can also be utilised for operational control following further analysis. The Forestry Commission's payroll is based on a weekly cycle with most payments being made in cash. Hence this part of the administration function is time-critical. Each Forest District is responsible for the administration of its own payroll.

## SYSTEMS DEVELOPMENT AND SOFTWARE SELECTION

The classical approach to systems development has been for a user to provide a specification outlining requirements. The system analyst then produces a formal specification after detailed discussion with the user. This specification is eventually agreed (and usually signed) by the user as an exact representation of what is required. Finally software is written using one of the available languages and tested by the programmer. The system is then returned to the user who after further testing accepts it. This approach works well in theory but suffers from certain drawbacks.

Firstly the user may experience difficulties in explaining the subtleties involved to the systems analyst. The result of this is that the finished product may not be quite that required. Secondly, whilst the user may have a good idea of what is required at the outset, the finished system may provoke a desire to enhance the system to take advantage of unforeseen possibilities. This may result in additional systems being created. Finally the classical approach can be very time consuming in terms of systems development productivity. Equally there is a risk that amendments and enhancements can be so time consuming as to prohibit the work being done. Martin (1982, Chapter 4) gives a good account of these problems and explains their impact on the productivity of systems development.

During the development of the Forest District system a major area of concern was software selection. The following considerations were felt to be of major importance in selecting software:

a) A high level of hardware independence. The operating system and applications software should run on a large number of different makes of micro-computer. The objective was to avoid the need to recode programs should the system be used on different hardware to that used in pilot studies.

b) The availability of a software package (application generator) to enable systems prototyping as a means of quickly exploring the practical implications of proposed development.

c) The availability of software packages for certain functions such as payroll.

d) The availability of multi-user software incorporating record and file locking to control input and update.

These features lead to the choice of the MBOS/5 operating system (manufactured by BOS Software Limited). The operating system was developed during the late 1970s and is widely used in Government Departments. Hardware independence has been achieved by placing all the machine related codes in one block of the appropriate program. The applications software does not therefore need to be recorded to enable implementation on different makes of hardware. MBOS/5 was one of the first truely multi-user (as opposed to concurrent) micro-computer operating systems.

The range of software available from BOS includes a number of different applications packages. Two packages are used in the system that has currently been developed. These are the BOS Payroll package and a package called AutoClerk. The payroll package is used for the calculation of gross to net pay and for producing payslips and other associated printouts. AutoClerk is an applications generator which enables data records and files to be defined. Screens for the input of data can be quickly developed as can reports (both screen and paper copy). AutoClerk was extensively used during the systems prototyping phase.

A language compiler was used to develop programs that were beyond the scope of AutoClerk. The software associated with the compiler includes a number of development tools. One of these was the job management facility which was particularly useful in instances where pre-determined commands and responses have to be followed by the user.

The advent of the so-called fourth generation languages has helped to significantly reduce the difficulties associated with systems development. There is an increasing trend towards systems prototyping usually replacing or at least supplementing the specification stage. During development of the Forest District systems, AutoClerk was used to prototype input screens and output reports. This enabled a systems framework to be rapidly developed. Once the outside edges of the systems have been developed it was then possible to identify input and output relationships relatively easily. Relationships were expressed in terms of volumes of data and their frequency. Any mathematical relationships were also stated. Two further areas were also considered during systems development.

Firstly, the ranges of valid data that could be input were established. A frequent cause of error in many systems is that invalid data can often be input. The result of this is that reports output from the systems often have little value. The various fields on each input screen are therefore validated wherever possible by reference to look-up files or other data input from a different screen.

Secondly, every attempt was made to ensure that the system was user-friendly. This (often clichéd) term is defined by Peltu (1981 p192) as 'appealing to a user with no computer experience'. The maximum use was made of menus and the BOS menu facility allowed hierarchies of menus to be quickly developed if necessary.

In addition the AutoClerk form facility was used for data input so that the user only has to remember two instruction messages. Purpose written information messages are included, where appropriate, to help the user to determine the validity of data that are permitted.

### APPLICATIONS CURRENTLY AVAILABLE

The system currently developed provides management information in four main areas and is summarised in Figure 2.

a) wages expenditure (including calculation of gross to net pay);
b) operational performance of vehicles, machinery and equipment (VME);
c) stocks of timber produced by FC harvesting operations;
d) job costing (using data from a and b above).

Data summaries from each of these areas are provided for transfer to the Conservancy (Regional) level. These transfers are currently achieved by paper though developments to enable the transfer of data files over the public telephone network are currently underway with a view to introduction in late 1985/early 1986.

Figure 2 shows that three types of data input are involved. Reference data apply to all Forest Districts and include files containing valid account codes, species and product categories and so on. These files are provided at system installation and may occasionally be updated. Master data are specific to individual Forest Districts but do not change very frequently. Examples include annual budgets, VME fleet details, timber contract summaries and so on. These data are set up as and when required by

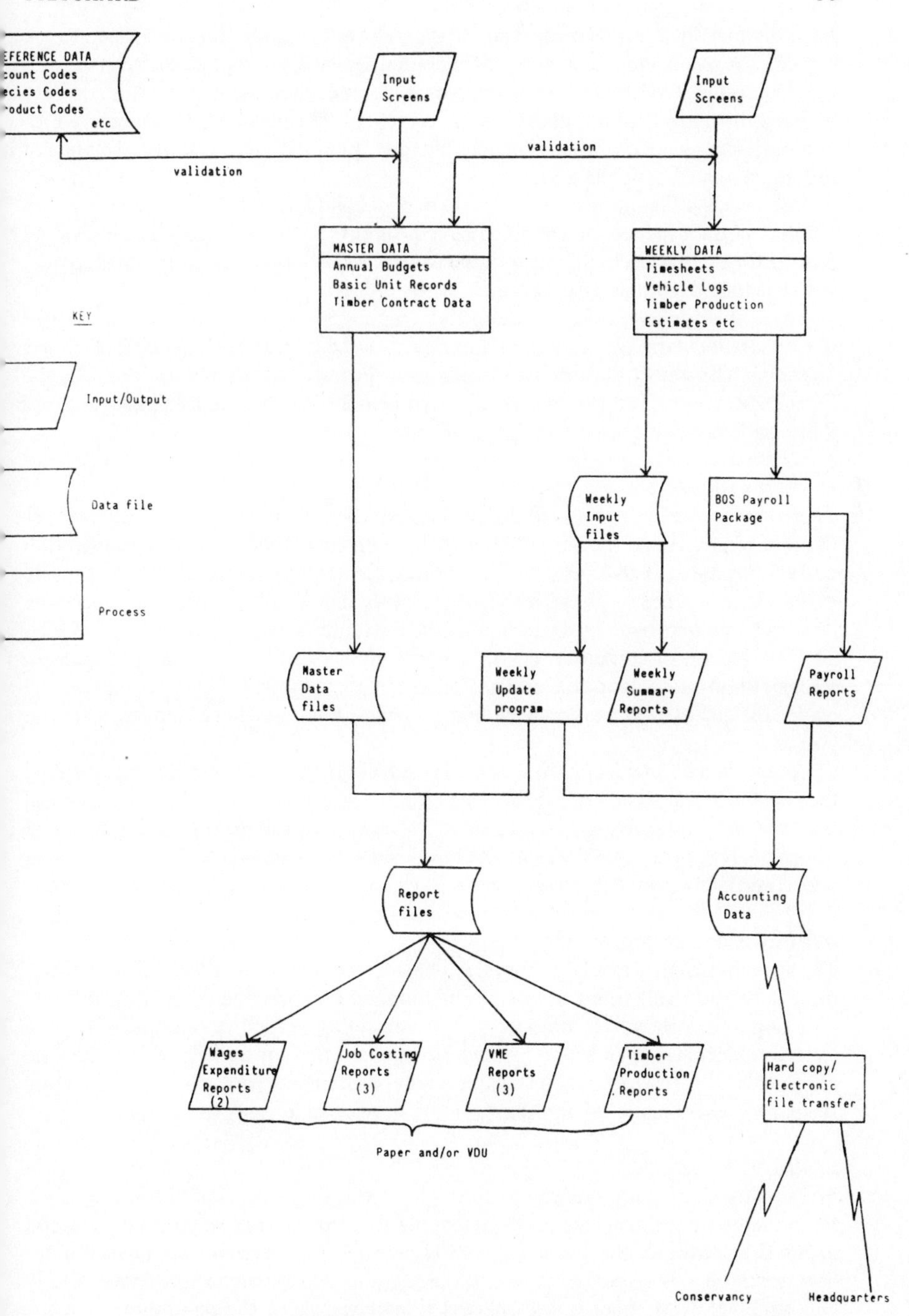


FIGURE 2. Forest district computer system summary.

the individual location. Weekly data are the basic information that are collected on a regular basis—usually on a cycle not exceeding one week. Employee timesheets and timber production estimates are examples. Reference data are used to validate data which in turn are used to validate the weekly data. The master files can be used for comparisons of actual performance against planned performance. Important differences can be highlighted.

Each week a file-updating process takes place and the weekly data are written to the master data files to enable management reports to be produced. The individual reports for a given week can be produced as required at any time until the next update. Performance to date is also included on the reports.

There are four categories of management reports as well as summaries for transfer to Conservancy and Headquarters. The reports available range from detailed reports on each job for use by the individual supervisor (Forester) to summaries for use by the District Manager. The summaries highlight broad performance and indicate those jobs which are performing notably different to expectation.

BENEFITS OF SYSTEM

There are three main benefits of the system. Firstly, the system significantly reduces the cost of processing the basic data in the management reports. The administration cost of the payroll is also substantially reduced. Secondly, the availability of accurate management information at frequent intervals enables the quality of operation control to be improved by responding quickly to abnormal or unpredicted events. Thirdly, the use of computer technology makes it possible to provide management information that cannot be cost-justified in a clerical system. This information will enable the quality of operational planning to be improved and hence reduce operating costs.

An internal financial appraisal of the system assuming a 10-year life and including the cost of full hardware replacement (to take advantage of technological advancement) after five years suggests that the system will yield an internal rate of return equivalent to eight percent. The appraisal also includes costs for staff training and conversion from the present clerical systems to the micro-computer system.

OTHER DEVELOPMENTS

The system currently available provides a base for further development. For example, there is considerable scope to extend the range of components to meet operational planning requirements. Components are currently under development by the Forestry Commission's Planning and Economic Division to help with economic appraisals of forest operations. Additionally, systems to meet the needs of more specialised users (such as nurseries) will be developed in response to demand.

ACKNOWLEDGEMENT

Thanks are due to my colleagues throughout the Forestry Commission for their many suggestions during the development of the system. I am also especially grateful to Mr E.K. Arthurs, the Forestry Commission Data Processing Manager, for his comments on this paper. Finally, I should like to express my appreciation to my colleague Mr R.M. Spence who played a leading role in the development of the system.

REFERENCES

Forestry Commission (1983) Report of the Forest District ADP Working Party. Unpublished Forestry Commission Internal Report.
Martin, J. (1982) *Application Development without Programmers*. Prentice Hall, Englewood Cliffs, New Jersey.
Peltu, M. (1981) *Using Computers—A Managers Guide*. NCC Publications, Manchester.
Pritchard, M.A. (1984) The Forest District Computer System—An Overview. Unpublished Forestry Commission Internal Report.

# OPERATIONAL PLANNING IN BRITISH FORESTRY
## A Group of Programs for the small user

J.D.A. WILLIAMSON

*Badenoch Land Management Ltd., Alvie Estate Office, Kincraig, Kingussie, Inverness-shire, Scotland PH21 1NE.*

*Summary*
Specifications for a group of programs currently under development are outlined. They include a financial program suitable for annual forestry accounting, a program for storing and accumulating physical and financial stand data, programs for volume estimation, growth models for predicting future stand production for even-aged coniferous stands in Great Britain and a financial comparison that can calculate the net present worth of a stand under a given management regime from data provided by the relevant growth model.

*Introduction*
The forest manager and woodland owner are interested in planning for two closely linked products of a forest enterprise: timber volume and revenue. The programs described here assume that the forest manager will plan to maximise one or other of these outputs given various constraints. These programs do not choose the best management option they only predict the physical and financial consequences of a given decision. The information required for these programs can be categorised as physical, financial and operational input data.

The small user in this context is the manager of a forest enterprise who owns or has access to a microcomputer. The forest enterprise may be one activity on an estate or it may stand on its own as a separate business. The small user generally has limited financial resources and the cost of collecting data for management purposes must be justified by the value of the decisions to be made for the enterprise.

A prerequisite in any planning is to know exactly what is on the ground. The stand is used as the basic topographical unit with one or more stands forming a compartment and several compartments making up a forest. The stand's physical inventory must include species, age, a measure of spacing and a measure of growth. Financial data can be specific to a given stand, for example the cost of fencing or building roads for given stand, or it can be general, such as the cost per unit area for planting or ploughing. Where there is no specific cost or revenue of an operation for a given stand, the cost or revenue can be estimated by taking a general cost and allocating it in proportion to the stand being considered. Operational data are the proposed management decisions, what to do, when and how.

The computer has the ability to calculate, accumulate, summarise and categorise this information. In the short term these summaries of physical and financial information are required to plan daily operations, budget for labour and materials, and estimate production by market categories. The computer can then translate the

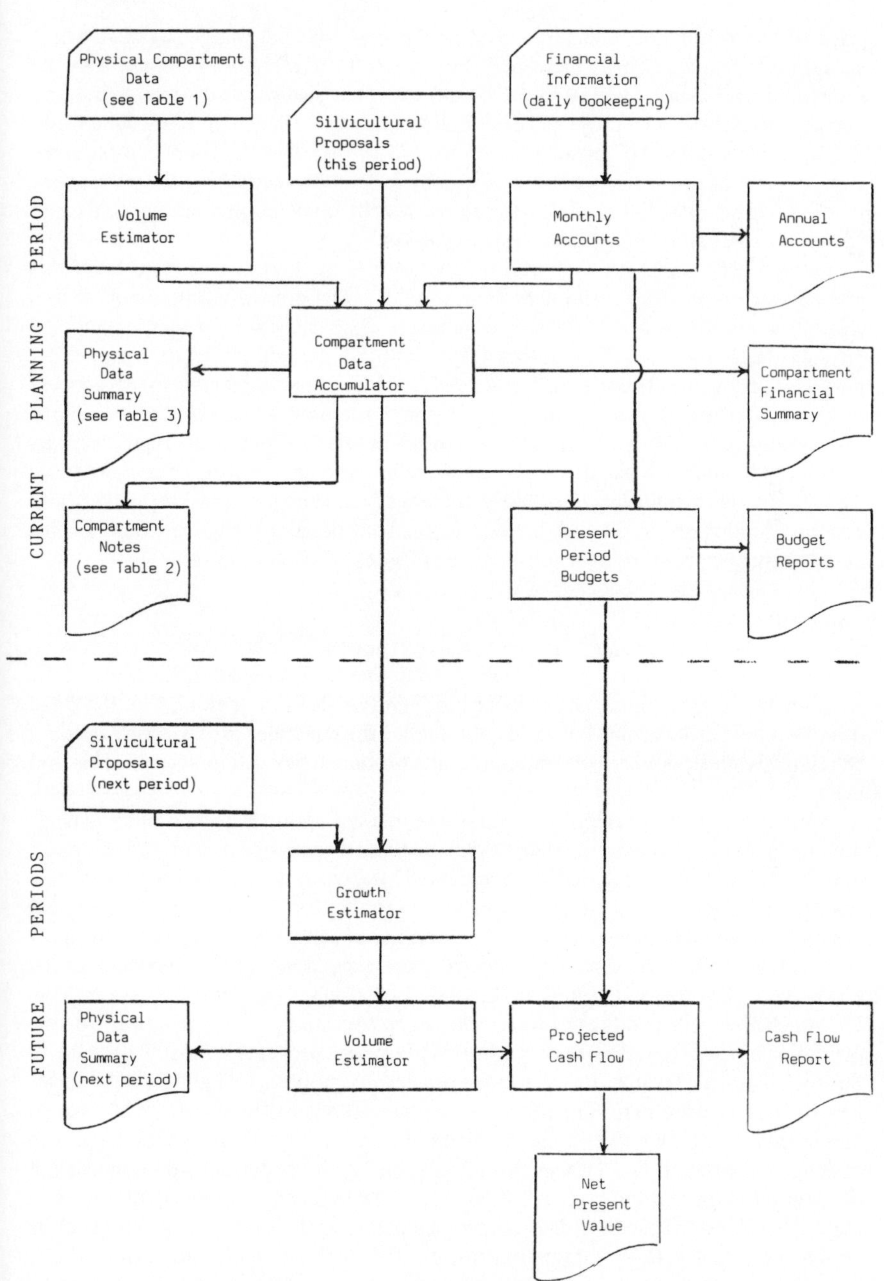


FIGURE 1. Outline of the interrelationship between programmes.

physical data into financial terms and produce a financial budget against which actual performance can be measured. For longer term planning of operations the forest manager usually wishes to predict the expected yields for any given period by market categories and a projected cashflow for the forest. When deciding which land to buy, which species to plant or when to thin or clearfell, the predicted financial consequence of the various options is usually judged by comparing the net present worth of each one. Estimating net present worth involves projecting yields and cashflows over as much as a complete rotation.

This process starts with what is on the ground at the moment, and from this predicts future physical stand data from the relevant Forestry Commission Tables (Hamilton and Christie, 1971), estimating future financial returns from the predicted physical data and then discounting these financial predictions to arrive at a net present worth value. There is sufficient relevant information to carry out this process in Britain today, but unfortunately the time required to make the necessary calculations is such that all but the most conscientious foresters make their decisions on a 'rule of thumb' basis. If we just had a quick and easy means of making these calculations we would be more likely to collect relevant information, make more accurate predictions and hence better management decisions. Figure 1 outlines this decision making process split into a series of inter-related programs.

### *Choosing a System*

Collecting data costs money. For the small user the cost of data collection is governed by the value of the management decision to be made. The smaller the value of the resource the less the manager can afford to spend on decision making. For the small user the required data must be readily obtainable, the machine for processing the data be relatively cheap and simple and the results produced be both understandable and relevant.

Agriculture and forestry in Britain are related industries with very similar management problems. Often the same management team has to make decisions on both land uses. For this reason we have treated forestry as a type of agriculture with a few unique problems of its own rather than try to develop a completely independent system that would only be relevant to forestry. In our case we have started with one of the currently popular suite of software programs designed for farmers called FARMPLAN. Farmplan Computer Services Ltd., from Netherton, Ross-on-Wye, Herefordshire have produced financial management and payroll programs for the farming industry together with various 'enterprise' programs mostly specific to farming. These programs run on a range of micro-computers including the Apple IIe which we have used here. The operating system used is called Programplan, which has certain advantages over 'Basic' in terms of speed, data storage capacity and ease of program maintenance. However, as far as the user is concerned, no knowledge of the computer operating system is required. The financial management program tackles the standard accounting problems peculiar to agriculture and forestry such as appreciating assets, the valuation of growing stock still on the ground at the end of a financial period and capital assets changing into units of production.

### *Annual Financial Program*

Income and expenditure for the forest enterprise is treated in the standard way for yearly accounting purposes. However, woodland accounts are peculiar in that the

TABLE 1. Compartment survey data.

| CPT No. | STAND No. | AREA (ha) | SPEC -IES | No. of TREES /HA. | PLANT -ING YEAR | TOP HEIGHT (m) | YIELD CLASS | MEAN D.B.H (cms) | BASAL AREA (m2/ha) | AV.VOL /TREE (m3) |
|---|---|---|---|---|---|---|---|---|---|---|
| 1 | 1 | 22.10 | SP | 3086 | 1976 | 1.5 | 6 | | | |
| | 2 | 8.50 | SP | 184 | 1869 | 18 | 4 | 28 | | 0.51 |
| | 3 | 0.90 | UP | | | | | | | |
| 2 | 1 | 27.00 | SP | | 1959 | 7 | 6 | 11 | 40.9 | |
| | 2 | 2.40 | OP | | | | | | | |
| 3 | 1 | 21.50 | NS | 3086 | 1985 | | 8 | | | |
| | 2 | 0.50 | UP | | | | | | | |
| 4 | 1 | 19.50 | SP | 233 | 1872 | 20 | 6 | 40 | | 1.07 |
| | 2 | | EL | 77 | 1872 | 21 | 6 | 41 | | 1.13 |
| | 3 | 0.90 | OP | | | | | | | |
| 5 | 1 | 8.30 | SP | 320 | 1870 | 21 | 6 | 33 | | 0.80 |
| | 2 | | EL | 17 | 1870 | 21 | 6 | 33 | | 0.80 |
| | 3 | 1.00 | OP | | | | | | | |
| 6 | 1 | 4.30 | SP | 997 | 1945 | 15 | 10 | 21 | | 0.25 |
| | 2 | 9.70 | SP | 3814 | 1964 | 4 | 6 | 5 | | |
| | 3 | 2.10 | UP | | | | | | | |
| 7 | 1 | 8.10 | SP | 317 | 1871 | 22.5 | 8 | 39 | | 1.17 |
| | 2 | 1.00 | OP | | | | | | | |
| | 3 | 0.10 | UP | | | | | | | |
| 8 | 1 | 24.8 | SP | 309 | 1872 | 20 | 6 | 36 | | 0.90 |
| | 2 | 2.00 | UP | | | | | | | |
| | 3 | 0.40 | OP | | | | | | | |
| 9 | 1 | 12.50 | SP | 3000 | 1984 | | 8 | | | |
| | 2 | 3.40 | NS | 3000 | 1984 | | 8 | | | |
| | 3 | 1.30 | UP | | | | | | | |

TOTAL AREA 182.30 ha

Species code SP = Scots Pine
EL = European Larch
HL = Hybrid Larch
NS = Norway Spruce

OP = Open but plantable
UP = Unplantable

annual profit and loss account is very sensitive to the way in which the growing timber is valued at the beginning and end of each financial year. In this program each forestry compartment or age-class can be considered separately. There are several possible methods of valuing the growing crop.

i) To take an assessed market value for each compartment or group of compartments as the opening value, estimate the felled value at the end of the rotation and calculate an annual value increment with which to increase the opening value to arrive at a closing value. This method will show a loss in years of high silvicultural activity and profits in years when no work is carried out.

ii) Calculate the closing value as the opening value plus the net expenditure for the year on that particular compartment or group of compartments. This produces an

TABLE 2. Example of compartment notes this period.

COSTINGS FOR THE PURPOSE OF ILLUSTRATION ONLY.

COMPARTMENT : 1 NAME : DUNACHTON MARCH TAX SCHEDULE B
GRANT BASIS 2

| STAND No. | AREA (ha) | SPEC -IES | No. of TREES /HA. | PLANT -ING YEAR | TOP HEIGHT (m) | YIELD CLASS | MEAN D.B.H (cms) | BASAL AREA (m2/ha) | AV.VOL /TREE (m3) | TOTAL VOL. (m3) |
|---|---|---|---|---|---|---|---|---|---|---|
| 1 | 22.10 | SP | 3086 | 1976 | 1.5 | 6 | | | | |
| 2 | 8.50 | SP | 184 | 1869 | 18 | 4 | 28 | 35.59 | 0.51 | 797.64 |
| 3 | 0.90 | UP | | | | | | | | |
| TOTAL | 31.50 | | 3270 | | | | | | | 797.64 |

PRESCRIPTION FOR PERIOD

| YEAR | DESCRIPTION OF OPERATIION | ESTIMATED EXPENDITURE | ESTIMATED INCOME | ESTIMATED BALANCE |
|---|---|---|---|---|
| 1984 | MANAGEMENT GRANT | | 66.71 | |
| | REMOVE DEER | 60.00 | | |
| | FENCE REPAIRS | 70.00 | | |
| | FIRE INSURANCE | 15.28 | | |
| | 1984 BALANCE | | | -78.57 |
| 1985 | MANAGEMENT GRANT | | 66.71 | |
| | REPAIR ROAD | 120.00 | | |
| | FIRE INSURANCE | 17.17 | | |
| | 1985 BALANCE | | | -70.46 |
| 1986 | MANAGEMENT GRANT | | 66.71 | |
| | FIRE INSURANCE | 18.11 | | |
| | 1986 BALANCE | | | 48.60 |
| 1987 | MANAGEMENT GRANT | | 66.71 | |
| | FIRE INSURANCE | 19.06 | | |
| | 1987 BALANCE | | | 47.65 |
| 1988 | MANAGEMENT GRANT | | 66.71 | |
| | FELL ST.2 | | 9576.00 | |
| | FIRE INSURANCE | 19.96 | | |
| | 1988 BALANCE | | | 9622.75 |

accumulation of historical costs for the growing crop and bypasses the need to estimate the crop's physical growth.

iii) Treat the growing timber as a capital item with nil or negative depreciation. The negative 'depreciation rate' being the estimated average annual rate of volume growth translated into financial terms. Silvicultural operations can be treated as capital improvements.

The manager can choose the method to be used; he will to a large extent be governed by the amount of physical and financial information available to him. The option chosen will have to be applied consistently throughout the forest enterprise.

Monthly budgets and cashflow projections can be made with reference to previous experience and future management plans. The manager can use the volume estimating and compartment data accumulating programs to predict the resultant cashflow given various management decisions. This forms the basis of the annual financial budget which is fed into the financial management program on a monthly basis to produce a projected cashflow for the following year. Actual results are input in the normal way and actual financial performance measured against this budget on a monthly and a cumulative basis for the year. By having income and expenditure split by compartments, the monthly cash effect on the business of proposed operations for any given compartment can be assessed. The value of the computer lies in its ability to predict the overall effect of a management decision in financial terms before the operation is carried out.

*Compartment Data Accumulator and Volume Estimation*

Table 1 gives an example of the information that is likely to be available from a management survey for stands of different ages. For the purpose of prediction it is important to enter a yield class even if it is only a guess as has been done for compartment 3 stand 1 in Table 1.

Given the information presented in Table 1, merchantable stand volume is normally estimated by looking up in the appropriate Management Tables (Hamilton & Christie, 1971). However, it is also possible for the program to make its own estimate of merchantable stand volume using this information for species where the stand volume, basal area, age, height relationship have been described in the form of equations. These are adaptations of main crop form height equations produced by Christie (1970). For more accurate volume estimates the British convention is to use Tariff Tables (Hamilton, 1975) and a separate program is being prepared to produce the missing figures from Table 1 for stands where sufficient data are available for the program to use the Tariff method of volume estimation.

The information produced so far is only of limited use to the forestry manager. What he really wants to know is the expected volume production by species and market categories and the net income from the proposed operations. The felled value of each compartment can be estimated from the Table 1 information by using stand volume assortment Tables (Hamilton, 1975) and market prices. Stand volume assortment tables can be translated into a series of equations, one for each merchantable category as described by a minimum top diameter and minimum length. By means of these equations the program can split an estimated volume into market categories (see Table 3). In the case of stands whose volume has been estimated by the Tariff method, single tree volume assortment tables (Hamilton, 1975) are used by the program for each diameter class, thereby giving a more accurate breakdown into size categories than using stand assortment tables.

The forest manager now knows the merchantable volume of the forest, split by market categories. Having decided on what stands to clear fell and the number of trees, volume, or basal area to remove in thinnings from the stand being thinned, the program has the means of estimating the forest's yield by market categories, and the physical stand parameters following the proposed operations. If financial information is also input for each compartment or specific operation, financial yield and forest values can be estimated for the period which can be used in conjunction with the financial program to produce a financial budget for the period.

Of course all of the calculations described so far can be done manually with the help of the appropriate management tables. The advantage of using the computer is in

TABLE 3. Summary of timber production this period.

| YEAR | CPT. | STAND | AREA (ha) | MERCHANTABLE VOL.O.B.M3 TOP DIAMETER: MIN.LENGTH: 7cms 3m | 12cms 4m | 18cms 4m |
|---|---|---|---|---|---|---|
| 1984 | 5 | 1 | 8.3 | 314 | 112 | 1814 |
| | | 2 | | 14 | 30 | 678 |
| 1984 TOTAL | | | 8.3 | 328 | 142 | 2492 |
| 1985 | | | | | | |
| 1985 TOTAL | | | 0 | 0 | 0 | 0 |
| 1986 | 4 | 1 | 19.4 | 305 | 286 | 4269 |
| | | 2 | | 96 | 197 | 1394 |
| 1986 TOTAL | | | 19.4 | 401 | 483 | 5663 |
| 1987 | 6 | 1 | 4.3 | 10 | 81 | 72 |
| | 7 | 1 | 8.1 | 362 | 210 | 2430 |
| | 8 | 1 | 24.8 | 828 | 484 | 5591 |
| 1987 TOTAL | | | 37.2 | 1200 | 775 | 8093 |
| 1988 | 1 | 2 | 8.5 | 6 | 146 | 646 |
| 1988 TOTAL | | | 8.5 | 6 | 146 | 646 |
| PERIOD TOTAL | | | 64.9 | 1935 | 1546 | 16894 |

the time saved. The computer can summarise the results from each compartment, produce the information for Plans of Operations as required by the Forestry Commission in the appropriate format, and allow the forest manager the time and means to adjust the proposed operations to produce a desired end result.

*Growth Prediction*

Many growth models have been produced but few are in a readily usable form for the small scale practising forester. (Hepp, 1982). Growth and species differ throughout the world which means that growth models are only applicable to the area and tree species from which the data were collected. Edwards and Christie (1982) have produced comprehensive yield tables for most of the commercially grown tree species in Britain but Lowe (1983) in his literature search found a shortage of suitable equations for predicting tree growth in Britain and ended up using polynomial growth equations developed by Williamson (1976). Whether the program predicts future yields from polynomial equations or from searching a table stored within the computer's memory, the results as far as the program user is concerned should be the same.

Given the current physical data stored for each compartment by the compartment

TABLE 4. Example of financial summary for three periods.

COSTINGS FOR THE PURPOSE OF ILLUSTRATION ONLY.

| CPT. | AREA (ha) | JOB DESCRIPTION | PLANNING YEAR 1984 | 1985 | 1986 | 1987 | 1988 | 1989 1993 | 1994 1998 |
|---|---|---|---|---|---|---|---|---|---|
| 1 | 31.50 | MANAGEMENT GRANT | 67 | 67 | 67 | 67 | 67 | 334 | 334 |
| | | VERMIN CONTROL | -60 | | | | | -200 | -200 |
| | | FENCE REPAIRS | -70 | | | | | | |
| | | FIRE INSURANCE | -15 | -17 | -18 | -19 | -20 | -114 | -52 |
| | | REPAIR ROAD | | -120 | | | | | -120 |
| | | FELL ST.2 | | | | | 9576 | | |
| | | PLANT | | | | | | -2805 | |
| 2 | 29.40 | MANAGEMENT GRANT | 64 | 64 | 64 | 64 | 64 | 320 | 320 |
| | | FIRE INSURANCE | -44.86 | | | | | | |
| | | REPAIR ROAD | | | | | -200 | | |
| | | THIN ST.1 | | | | | | 675 | |
| 3 | 22.00 | MANAGEMENT GRANT | 47 | 47 | 47 | 47 | 47 | 234 | 234 |
| | | FENCE | -3500 | | | | | | |
| | | VERMIN CONTROL | -200 | -40 | -40 | -40 | -40 | -200 | -200 |
| | | PLANT | | -6450 | -690 | | | | |
| | | PLANTING GRANT | | 2257 | | | | | |
| | | FIRE INSURANCE | | -5 | -6 | -7 | -8 | -52 | -80 |
| | | FENCE REPAIRS | | | | | | -200 | -200 |
| 4 | 20.40 | MANAGEMENT GRANT | 44 | 44 | 44 | 44 | 44 | 220 | 220 |
| | | CLEAR FELL | | | 79282 | | | | |
| | | FENCE | | | | -6800 | | | |
| | | VERMIN CONTROL | | | | -150 | -40 | -200 | -200 |
| | | PLANT | | | | | -6120 | -650 | |
| | | PLANTING GRANT | | | | | 2142 | | |
| | | FIRE INSURANCE | | | | | -5 | -36 | -52 |
| | | FENCE REPAIRS | | | | | | -200 | -200 |
| 5 | 9.30 | MANAGEMENT GRANT | 20 | 20 | 20 | 20 | 20 | 100 | 100 |
| | | CLEAR FELL | 34888 | | | | | | |
| | | FENCE | | -3800 | | | | | |
| | | VERMIN CONTROL | | | -110 | -20 | -20 | -100 | -100 |
| | | PLANT | | | -2790 | | | | |
| | | PLANTING GRANT | | | 976 | | | | |
| | | FIRE INSURANCE | | | -2 | -3 | -3 | -23 | -35 |
| | | FENCE REPAIRS | | | | | | -100 | -100 |
| | 112.60 | TOTAL CARRY FORWARD | 31240 | -7933 | 76844 | -6797 | 5504 | -2997 | -331 |

data accumulator and estimated by the volume estimating program, a compartment's physical data can be predicted for the next planning period. Ideally the program should be able to reproduce for the next planning period most of the compartment physical and financial data already estimated for this planning period (see Tables 2 and 3). The process can then be repeated for further planning periods to build up a prediction of both timber production and cashflows as shown in Table 4, given a particular set of management decisions in the present planning period.

*Present Worth Calculator*

Most forest owners wish to know the estimated net present worth of their investment. The net present worth of the stand can be calculated for different management decisions by repeating the growth model for a given stand from planting to clear felling and discounting back the resultant cashflow. Thus the investor can decide whether or not to plant an area of similar growth potential to planted ground already monitored, or alternatively to sell off a particular area. For predicting the net present worth of the forest one can estimate the present market value, predict the growth and cashflow for several periods into the future and discount back the predicted cashflow and future market value for the end of the simulation period. Add in the effects of tax concessions and the forester has a useful tool for persuading clients to invest in a given forest proposal.

*Conclusion*

Once information is stored on computer it can be quickly retrieved, sorted, accumulated and summarised as required. So far the British forestry manager lacks the specialist management software which is already available in other industries. Hopefully this situation will change over the next two or three years. Many of the relationships between stand parameters required for forestry management programs are not formulated in mathematical equations readily adaptable for use by a micro-computer. However, it is unlikely that anyone is going to spend time and money adapting these relationships until there is software available to use them. The development of software for forestry must be considered like the development of the motor car. The first models are and will be crude, but until the first model is running there is no base on which to build.

REFERENCES

Christie, J.M. (1970) *The Characterisation of the Relationships between Basic Parameters in Yield Table Construction.* Paper presented Sept. 7-11 IUFRO advisory group of Forest Statisticians, Jony-en-Josae, France.

Edwards, P.N. and Christie, J.M. (1982) *Yield Models For Forest Management.* Forestry Commission booklet no.48, H.M.S.O., London.

Hamilton, G.J. (1975) *Forest Mensuration Handbook.* Forestry Commission booklet no.39, HMSO, London.

Hamilton, G.J. and Christie, J.M. (1971) *Forest Management Tables (Metric)* Forestry Commission booklet no.34, HMSO, London.

Hepp, T.E. (1982) YIELD: Timber Yield Forecasting and Planning Tool, *Southern Journal Of Applied Forestry,* 135-140.

Lowe, P. (1983) *The Use of Microcomputers for Yield Prediction.* Unpublished paper Dept. of Forestry, Univ. of Aberdeen.

Williamson, J.D.A. (1976) *Decision Making in British Forest Management for the Private Woodland Owner by Means of Simulation and Integer Programming.* Unpubl. PhD thesis, Dept. of Forestry, Univ. of Aberdeen.

# A MODERN FORESTRY PLANNING SYSTEM

S. WESTERLING

*Skogsarbeten, Box 1184, S-163, 13 Sparga, Sweden.*

## SUMMARY

This paper describes a new approach to forestry planning. It consists of an integrated planning system with routines covering all current forestry plans, from strategic to short-term ones. The system contains a number of computer-based routines. The user, however, has the option of replacing some with manual ones.

The planning system is best described through the following four italicised keywords.

It is *decentralised,* since it can be used throughout the organisation, either by way of terminals or micro-computers.

Everyday language is used throughout and necessary commands are entered with the help of menus or similar devices. In this way the system satisfies the demand for *user-friendliness.*

No actual planning is carried out by the computer. It merely helps the planner to keep track and to perform calculations. In that sense the system is *decision-supporting.*

The system is at least in a broad sense *independent* of the make of computer and the type of data available.

## INTRODUCTION

Skogsarbeten (The Forest Operations Institute of Sweden) has for a number of years been working on the development of co-ordinated planning systems for use in forestry. Most of the work has been carried out on a consultancy basis for individual organisations, but the routines developed are of a general nature and require only limited modification before being suitable for use by any forest enterprise.

There are two basic reasons for this development work. The first and main one is that co-ordinated systems in use for planning were almost non-existent in the Swedish forestry sector (Ericson and Westerling, 1981). The routines that did exist were not related to each other and, consequently, were often used in isolation. The other reason is the steady increase over the years in the size of the forest area for which individual managers are responsible. In the Swedish large-scale forestry sector the value of such a forest area could range between 50 and 80 million Swedish kroner. This has made it difficult, if not impossible, for the person concerned to do the planning in his head.

The intention of this paper is not to give a detailed descrption of the routines used in the planning system Skog Plan -84. It will, instead, deal with some of the more important features and advantages of the system from the viewpoint of the user. Before going into that, however, it is necessary to give a brief description of the system in general terms.

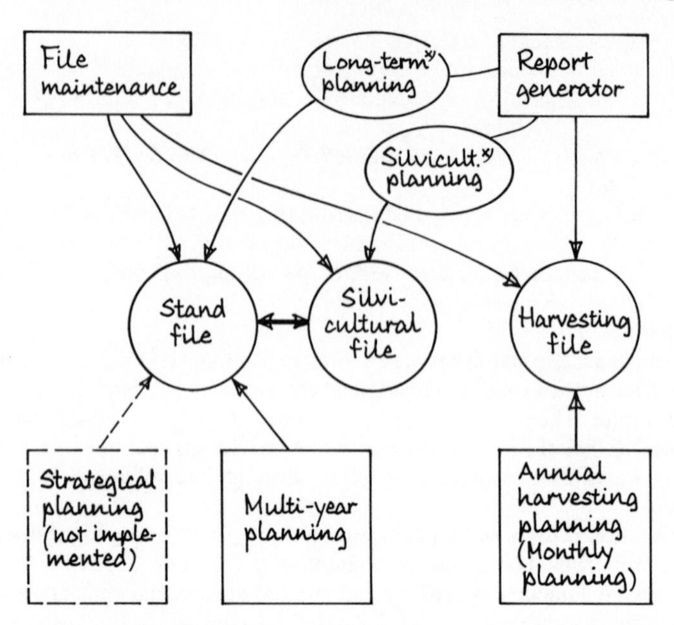


FIGURE 1. Overview of the planning system.

*No special procedure, done with help of the report generator.

| | Strategical Planning | Planning of harvesting and silviculture | | | |
|---|---|---|---|---|---|
| | | Long-term | Multi-year | One-year | One-month |
| Time-span | 50-100 years | 10-15 years | 3-5 years | 1-1½ years | 1 day-3 months |
| Types of decision | Volume to cut per 10-year period | | (Local level) | | Scheduling |
| | Cutting policy Thinning program New species? | Distribution between final felling and thinning. District. Type of stands to be treated. | Which stands with regard to concentra-tion, losses of growth etc Treatment year for each stand. Road building | Time of the year for treatment. Operating method. Resources/season | Which team for a given stand. |

FIGURE 2. Planning procedures in Sweden

The main components of the system are shown in Figure 1. As can be seen, the system includes procedures for both long- and short-term planning of harvesting and silviculture. The procedures are computerised but those for a shorter time-span, can be replaced by manual ones.

The scope of the forestry plans used in Swedish forestry are shown in Figure 2.

Multi-year plans and those with a longer time-span are based on a stand file. This file contains the organisation's total forest holdings broken down into smaller units—stands—with one case to each stand. Each stand is normally described through the following groups of variables:

Identity
Compartment data (site-index, area, etc)
Growing stock (volume, basal area, age, etc)
Treatment suggestions

It should be mentioned that the data in the stand file are a result of a fairly crude inventory and consequently are not always accurate.

Annual and monthly planning, on the other hand, are based on files (harvesting and silviculture) containing more detailed data. The units in the silvicultural file are the same as in the stand file (see the connection in Figure 1), but due to the speed at which young stands develop it is necessary to store information in more detail for these particular stands.

The silvicultural units are created in a new inventory shortly after clear-felling. If necessary, the area is divided into stands according to company policy and each stand is also described in the stand file. The treatment of each unit is planned in the field. Both planned and performed treatments are stored in the silvicultural file. The following groups of variables are stored:

Identity
Site preparation (method, area, date etc)
Reforestation (planting method, material etc)
Cleaning (number of stems before and after, species etc)

The silvicultural units are removed from the file when the next treatment in the life of the stand is first thinning.

The harvesting file contains data about harvesting units. These units are formed as a result of a new inventory. The list of stands to be included in the inventory is produced by the multi-year plan (see Figure 2). The inventory is now often placed on a quantitative basis and the data collected are consequently of a much higher quality. The harvesting units seldom, if ever, correspond with the stands mentioned earlier. The variables in the harvesting files are normally grouped as follows:

Identity
Site data (site-index, area, ground conditions, etc)
Growing stock (volume, species, quality, etc)

It is very important to point out that when drawing up the plans with the help of the computerised procedures, the user must also use other instruments such as local knowledge, maps, company policy, and so forth.

The system described has been in operation at a Swedish enterprise (Bergstrand, et al., 1982). A new and improved version is now being developed in co-operation with another Swedish enterprise (Hargs Godsförvaltning AB). In the following discussion the figures used to describe events on the visual display are extremely simplified. This is done on purpose, since the aim is to demonstrate general principles.

### THE DECENTRALISED SYSTEM

If a planning system is to be used, it is necessary that the people who will use the plan in their daily work, and thus benefit from it, also play an active role in the planning process. They are often the only people within an organisation possessing the local, detailed knowledge which will make the final plan practical and not just a product of an armchair exercise.

It is necessary to have a formal mechanism for keeping the files up to date, as well as making use of the local knowledge. Since the main users (at least of the most recent data) are the people out in the field and fairly close to operations, it is natural that the same people are made responsible for the updating.

In addition to their use in formal planning, the data files are also used for 'unscheduled' retrieval of data. This requires the ability to answer, quickly and simply, questions such as: 'The weather is now extremely cold, where does the company have stands which have to be cut during extreme winter conditions?'

To satisfy these demands the planning system has been designed so that it can be used by people at the periphery of the organisations, either by means of micro-computers or terminals. It is important to point out that from the planning point of view there is no real need for terminal connections; mail is a good alternative.

### THE USER-FRIENDLY SYSTEM

An over-riding consideration during the design of the computerised functions of the planning system has been that a user without any computer background should be able to run the system. This has been accomplished by making the system user-friendly.

Everyday Swedish is used throughout the program. Great care has been taken to avoid messages from the operating system of the computer, since these messages are unfortunately normally in English.

Selections between subprograms and choices between different options are usually made with the help of menus. For longer specifications a so-called window feature is used as shown in Figure 3. The window is the area inside the box and it can, with the help of the keyboard, be used in the same way as paper, pencil and eraser. When the user presses the return key, only the displayed text will be processed.

Entering new units in a file is one of the features of file maintenance. The user enters the data in a window (see Figure 3) with the same format as the entries of the field sheet used during data collection. Consequently there is no need for space-consuming explanatory text. The data entered and the associated text are, however, displayed after verification, and errors discovered are therefore easy to correct. Since the data collection and keying in of data are usually done by the same individual, it is easy for the user to decide how errors should be dealt with.

The routines do not need to produce hard copy, since the screen is used both as note-pad and for displaying information. It is of course possible to produce hard copy for documentation, but this need is not great, since the tables are readily accessible.

A report generator is included in the system, which enables the user to produce tables of optional content and format. When developing the report generator, one aim

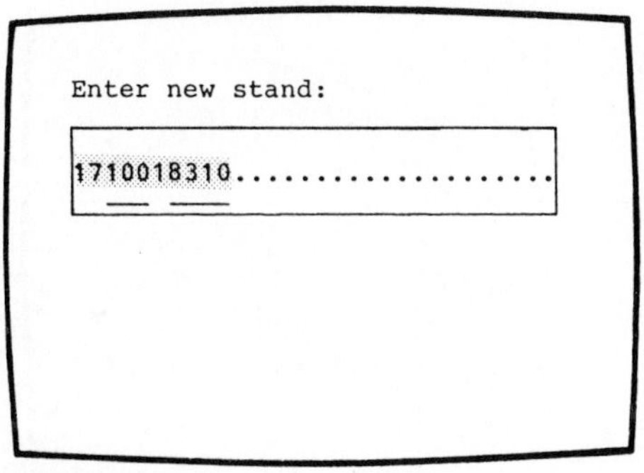

```
Enter new stand:

1710018310....................
```

*Explanation*

| | |
|---|---|
| Map no. | 17 |
| Stand no. | 100 |
| Labour area | 1 |
| Inventory year | 83 |
| Inventory month | 10 |
| etc. | |

FIGURE 3. Data entry of new stand Shadowed area entered by user.

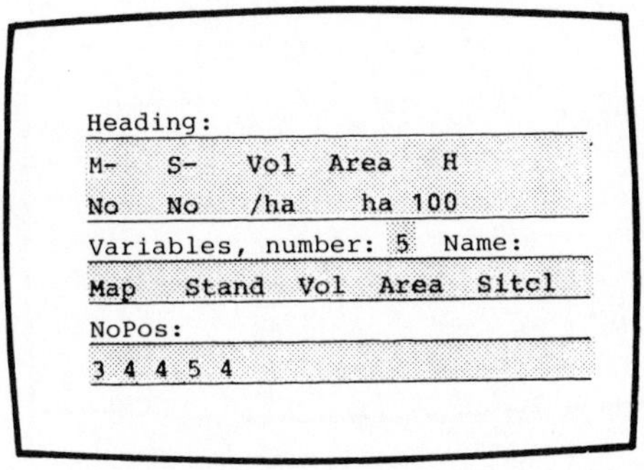

```
Heading:
M-   S-   Vol  Area   H
No   No   /ha    ha 100
Variables, number: 5  Name:
Map   Stand  Vol  Area  Sitcl
NoPos:
3 4 4 5 4
```

FIGURE 4. Example of table-specification Shadowed areas entered by the user.

was to keep the work of specifying the tables fairly simple. Nearly every table produced from files of the available type will have the following basic design:

Heading

One line per stand, unit etc

Totals

The report generator therefore produces such tables which is one way to simplify the report specification. Figure 4 shows an example where the user wishes to see a report covering five variables, namely, map and stand number, standing volume, stand area and site class. As can be seen, the only things the user has to worry about is the heading, the names of the variables that should be included and the number of positions for each variable (= each column). Another step towards simplicity was to decide once and for all how many decimals each variable should be presented with and how totals should be calculated. It is also possible to sort the contents of the table.

| M-No | S-No | Age tot | Area ha | H 100 | P |
|---|---|---|---|---|---|
| 1 | 371 | 85 | 4.0 | 120 | 0 |
| 1 | 372 | 97 | 5.1 | 126 | 2 |
| 1 | 380 | 112 | 8.2 | 228 | 3 |
| 2 | 145 | 148 | 3.7 | 218 | 3 |
| 2 | 192 | 119 | 2.1 | 220 | 3 |
| 2 | 207 | 90 | 4.8 | 116 | 0 |

Multi-year planning

*Explanation*
M-No = Map number
S-No = Stand number Identity
H100 = Site index
P = Priority rating entered by user. See text for further details

FIGURE 5. Example of basic selection—final felling Shadowed area entered by user.

Multi-year planning

| T-area | No st | Area ha | Cut m3/ha | m3f | P | Year |
|---|---|---|---|---|---|---|
| 1 | 13 | 65 | 162 | 8424 | 1.7 | 0 |
| 2 | 18 | 98 | 221 | 17326 | 2.3 | 2 |
| 3 | 12 | 53 | 197 | 8353 | 1.9 | 0 |
| 4 | 2 | 12 | 202 | 1939 | 3.0 | 1 |
| 5 | 16 | 102 | 241 | 19666 | 2.5 | 1 |
| 6 | 8 | 39 | 198 | 6178 | 1.9 | 0 |

*Explanation*
T.area = Treatment area
No st = Number of stands
$m^3$/ha = Volume per hectare (average)
$m^3$f = Solid volume (total)
P = Priority rating (average)
Year = Year of the period in which treatment area is harvested

FIGURE 6. Example of comparison between and allocation of year to treatment areas Shadowed area entered by user.

The minor limitations thus included do not create any problems for the user.

THE DECISION-SUPPORT SYSTEM

It is important to point out that though the planning in the system is carried out with the help of computerised procedures, the computer never performs any active planning work. The computer is merely the user's obedient servant carying out the onerous task of processing large quantities of data.

This can be illustrated by the multi-year planning procedure. It will also illustrate how the other routines are handled, as they work much in the same pattern.

The planning process starts off by extracting a framework for future activities from the long-term plans. This framework can, among other things, include areas suitable for final felling in the given time-span, selection criteria for thinnings and so forth. To this must be added details about available man-power and demands for wood, for example. The routine will then search the file and extract a basic selection of stands suitable for the treatment specified by the user (see Figure 5). The basis for selection are incorporated in the program and cannot be changed from one planning process to the other. Simple criteria are used, for instance in the case of final felling, stands older than a certain age are selected. In the case of reforestation, clear-cut areas are selected as well as stands selected for final felling during the planning process.

The planner then subjectively ranks each individual stand in the basic selection with the help of stand-data, local knowledge and maps, and other factors. In this example, three stands have been allocated high priority ranking (3) because of high age and fairly high site index (H100). By making decisions at the stand level it is possible to choose the most suitable time for treatment on biological grounds.

It is also desirable, at least in Scandinavia, to concentrate the activities in time and space. A new hierarchical level, called treatment areas (normally 50-100 stands), has for that purpose been introduced above stand level. The treatment areas could, for instance, consist of a number of stands accessible from the same road. The areas are marked on maps as well as included in the identity of the stand. The operation described in the preceding paragraph is in reality done once for each treatment area.

The planner takes the biological and economic considerations into account by comparing the *collated* stand particulars for the treatment areas, and not data from individual stands, when deciding whether a given treatment is to be carried out or not (see Figure 6). The multi-year plan normally covers a three-year period and the decision to treat a treatment area is made once the planner has allocated a specific year for the area. The aim is to treat as few areas as possible, but, when treated, every possible stand in the area shall be treated. The financial side is thus taken into account since the cost of moving equipment, keeping roads open in winter and managing labour can be minimised.

This is followed by a projection of the results over the next three years as illustrated in Figure 7. The projected results are compared with the production targets drawn up earlier and any necessary adjustments at stand or treatment area level are made. This is repeated until an acceptable result is obtained.

## THE INDEPENDENT SYSTEM

The system can basically be run on any size or make of computer. However, both the main and external memories have to be of a certain minimum size. Because it must be possible to retrieve information quickly, the computer must also have access to a hard disc.

A big problem with computer systems today is the portability of the software. The programs of the planning system are, however, fairly simple to adapt to different

Consequence analysis

| | *1* | *2* | *3* | year |
|---|---|---|---|---|
| Area | 124 | 98 | 103 | ha |
| Vol tim | 12963 | 9529 | 8713 | m3f |
| pul | 8642 | 7797 | 6280 | m3f |
| Dist-% | 79 21 | 43 57 | 83 17 | 1-3,3+ |
| Cutter | 120 | 97 | 80 | manday |
| Harvest | 380 | 360 | 382 | mchday |
| Dist-% | 83 17 | 48 52 | 78 22 | 1-3,3+ |

*Explanation*
Vol tim = Timber volume
pul = Pulp-wood volume
Dist-% = Distribution between different ground conditions

FIGURE 7. Example of projected results Shadowed area result of calculations.

computers or programming-language dialects. This has been proved by conversions already carried out. Since at least some of the programs have to be adapted to the needs of individual users due to other reasons (such as specific types of data in the stand file), some programming has to be performed anyway.

The complete planning system is developed on a micro-computer with the following specification:

Make and model: IBM PC XT
Internal memory: 256 kbyte
External memory: 10 Mbyte
Operating system: DOS 2.0
Programming language: Pascal/MT+ (Digital research)

The programs will shortly be available through Skogsarbeten on a commercial basis.

## CONCLUSION

In the early seventies several attempts were made to construct planning systems as described in this paper. They were based on central computers and often contained complicated mathematical solutions such as linear programming. These attempts failed, mostly because of the technology, since the time between submitting the deck of punched-cards to the computer and receiving the results was far too long.

What distinguishes this planning system from these previous approaches? This can be answered through two keywords—method and technology.

The planning methods used are well-proven and workable, often only translated or standardised from the corresponding manual ones to computerised versions, without any decision-making included in the process on the part of the computer. The methods are also easy to learn, which is important since many of the routines will not be in daily use.

As regards technology, the system can be run on a modern micro-computer, and with the help of visual displays most of the need for pad and pencil is eliminated.

For whom is this system designed? That question cannot be answered by giving any precise figures for minimum turnover, annual cuts or number of stands. It can, however, be pointed out that when planning work becomes laborious and difficult to manage using ledgers or similar manual devices, it is time to consider changing over. Since the cost of computer equipment continues to fall, the computer is coming within reach for smaller and smaller organisations. In those cases, using the computer for accounting and similar purposes in addition to forestry planning could also be considered.

REFERENCES

Bergstrand K-G et al. (1982) *A coordinated Planning and Follow-up System* (In Swedish, with a summary in English.)—Forskningsstiftelsen Skogsarbeten, Redogörelse nr 2 1982.

Ericson, O. and Westerling, S. (1981) *Forestry Planning: The Present Situation and Future Trends* (In Swedish, with a summary in English).—Forskningsstiftelsen Skogsarbeten, Redogörelse nr 3 1981.

# PART III

# COMPUTER-AIDED DEVELOPMENTS IN FOREST PLANNING

# THE USE OF DATABASE TECHNOLOGY IN FORESTRY

BERNARD M. DIAZ

*NERC Scientific Services, Holbrook House, Station Road, Swindon, Wilts SN1 1DE*

SUMMARY

Databases gather data together under a 'data design'. However, to use the data design, operations on the data need to be done. The nature of these operations and their effect on the data are themselves gathered together in an 'operational analysis'. The two together constitute 'data analysis' which is the complex and important first step in exploiting database technology. The data analysis is implemented using a database management system (DBMS) which is the hardware and software, and which can come from a large number of available technologies.

The author sees database technology as a whole making a considerable contribution to many aspects of forest management and exploitation. These benefits will accrue, providing care is exercised, when making the original decisions to implement the database, the data analysis procedure and the choice and implementation of the appropriate database management technology.

## INTRODUCTION AND APOLOGIA

Although a botanist I know next to nothing about forestry. I do however know a little about computers and databases and the management of data as a resource. I also know a lot about the 'politics' of managing data and database management systems, principally owing to my job advising the Natural Environment Research Council on its database strategy, but also because of the need to go and design and implement them. The intention of this paper is to describe how I would approach the problem of applying database technology to forestry.

## FIRST PRINCIPLES

Do you need them? What are the hassles? What is the cost?

Whether you need to have a database is a debatable question. The main guidelines are,

1. do you have *a lot* of valuable data,
2. are the *data needed* to help you do your business, that is, growing trees, managing people or money etc,
3. do you *need predictions,* which only a complex set of computations will allow you to reach,
4. are *many people* likely to need to access all these facts,
5. do you need *to impress* others, for whatever reason—politics, getting finance, selling your products, etc,
6. do you need them to *justify* the fun you get (you wish to get) out of computing.

If the answer to three or more of these questions is 'yes' (be honest), then you probably need to read the rest of this paper—and you may also need to use database technology.

The hassles are summed up by saying 'computers', 'jargon' and 'high priests'. I shall quote from 'Up The Organisation', an excellent little book I'd recommend to anyone embarking on the 'Management of Things' business. In it Robert Townsend (1970) says of computers:

> 'First get it through your heads that computers are big, expensive, fast, dumb adding-machine-typewriters. Then realise that most of the computer technicians that you are likely to meet or hire are complicators, not simplifiers. They are trying to make it look tough, not easy. They're building a mystique, a priesthood, their own mumbo-jumbo ritual to keep you from knowing what they—and you—are doing.'

Some rules of thumb:

> '. . . make sure the climate (for computer and database implementation) is ruthlessly hard-nosed . . .' ask 'what are you going to do with every (computer) report', 'what would you do if you didn't have it'. '. . . make sure your present way of doing it is reasonably clean and effective before you automate it. Otherwise your new computer will just speed up the mess'.

I will add nothing apart from noting that even micro-sized, fast, dumb, adding-machine typewriters can be a menace, and noting (in my defence) the words '*most* of the technicians . . .'. If you find a good and/or honest one, pay him well and lock him away.

Finally cost. A small database system can be implemented for as little as £5,000 and a recurring cost in the order of a £100 per annum. However, this assumes a single user (possibly on a personal computer) with modest aims and expectations. In this paper it is assumed that such an approach would be appropriate as an evaluation exercise but not for the full final implementation. At the other end of the cost scale—well, there is no other end to this scale.

## FIRST THOUGHTS

Having decided on using database technology what do you do about it? The first thing is to decide on the capital cost and recurring costs that you are willing to spend. It is likely these figures will float upwards, so be warned. The next is to get one A4 sheet of paper and write out the reasons for getting the system. On another sheet of A4 write out the objectives of the system as you see them. Put these two pieces of paper into a safe with your most valuable valuables. Next, decide on a timescale by which you would like to see the system working and hire someone to look at your current ways of doing things. Both these decisions will be constrained by your decision on costs. Spend the time waiting for the report in thinking about how you currently do things. When you get the report see if it confirms that current methods work tolerably effectively. If not, either hire new experts or think again, seriously, about the computation of the procedure.

These acts are the preliminary steps. What remains is where you enter priestdom, with one possible escape route. If you can, find a good and/or honest priest (see First Principles) then pay him to enter on your behalf. If this is impossible—read on.

To implement a database (the inter-related data) on a database management system or DBMS (the actual computer and its attendant software) you need to do the following:

FIGURE 1a

| Forest-name | Compartment number | Stand number | Stand type | Stand age | Number of trees |
|---|---|---|---|---|---|
| Calder | 1 | 3 | pine | 3 | 2000 |
| Knarerborough | 1 | 4 | spruce | 19 | 700 |
| Dean | 5 | 1 | oak | 120 | 95 |
| Grand | 11 | 5 | hazel | 12 | 75 |
| . | . | . | . | . | . |
| . | . | . | . | . | . |
| . | . | . | . | . | . |

FIGURE 1b

*Wood*

| Wood-type | Wood-age | Average volume |
|---|---|---|
| spruce | 80 | 100 |
| pine | 10 | 50 |
| hazel | 20 | 40 |
| oak. | 300 | 120 |
| . | . | . |
| . | . | . |
| . | . | . |

One fairly well understood data relationship is the table, in which there are columns, rows, column names, table names and items. Note that the table names and column names are vital pieces of data allowing you to find your way around the various relationships. Also note that within a column only one type of data is permitted. For example, only valid names of forests are allowed in the 'forest name' column in the STAND table and only positive integers in the 'average volume' column in the WOOD table.

1. a data design and operational analysis;
2. a survey of appropriate technology and choice of a DBMS;
3. the implementation and some system tuning;
4. to design evaluation procedures.

I shall attempt, in the remaining sections of this paper, to outline a little of what is involved in each of these areas.

## DATA DESIGN AND OPERATIONAL ANALYSIS

The data design for a database is the diagram which outlines the inter-relationships of the various data elements (Vetter and Maddison, 1981). For example, a well understood relationship is the 'table', which has 'rows' and 'columns' (examples are illustrated in Figures 1a and 1b). Another relationship is the hierarchy in which for example a 'forest' contains 'compartments' and 'stands', there being many stands in a compartment and many compartments in a forest (Figures 2a and 2b). The process of data design reconciles the natural units such as 'forests' and its natural hierarchy of 'compartments' and 'stands' into artificial units such as 'Forest Name', 'Compartment Numbers' and 'Stand Planting Date'.

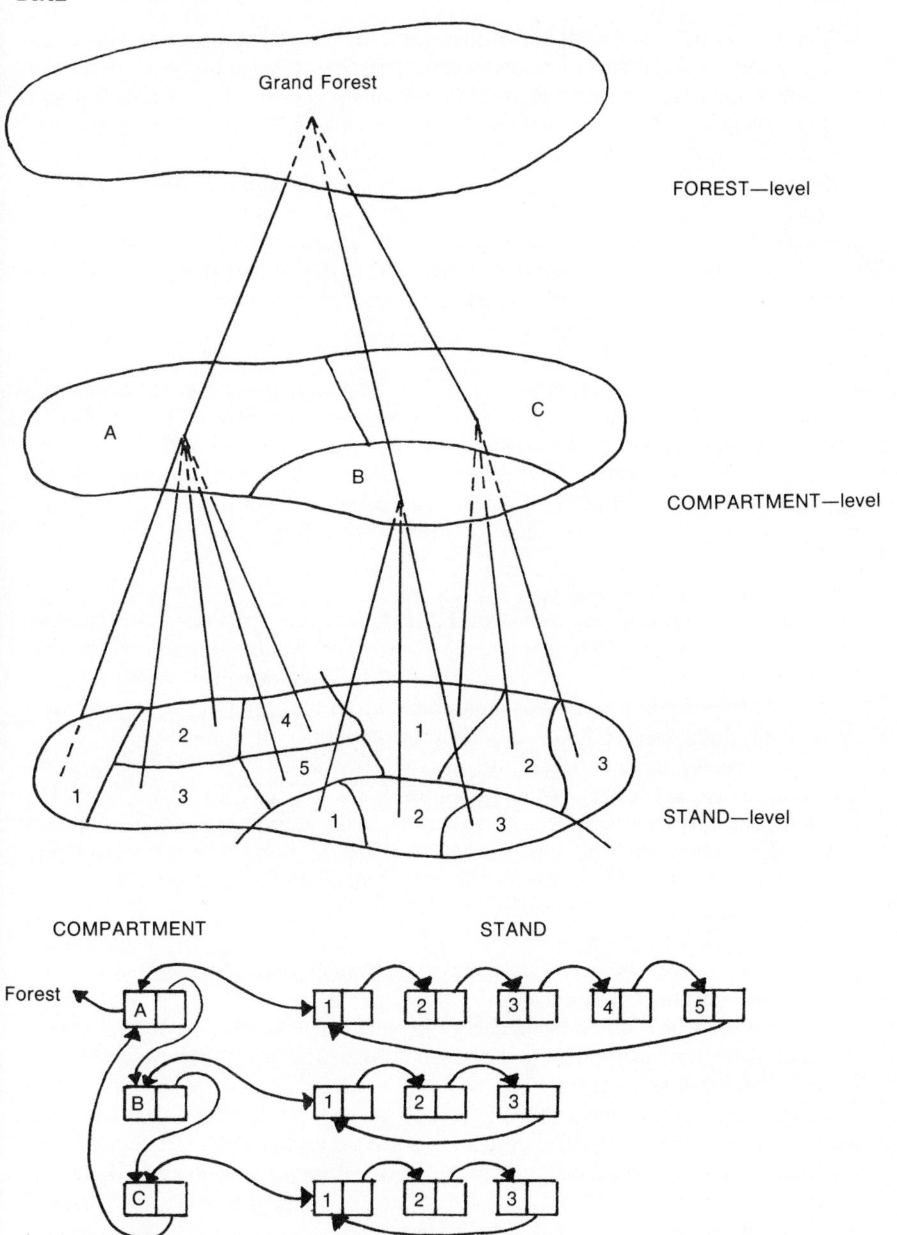


FIGURE 2a. An important and frequently occuring relationship is the hierarchy. This is illustrated in 2a with a forestry example. Other hierarchies can easily be brought to mind—think of a book which contains chapters, paragraphs etc. Many ways exist for putting such hierarchical data into a database. For example, four chains could be envisaged as in 2b, where each chain of STANDS is linked to an entry which is in a chain of COMPARTMENTS which itself is in a chain of FORESTS, and so on. (Be aware that simpler representations than this do exist . . .)

An operational analysis does two things. Firstly it assesses the operations which actually happen and ensures that they are reflected in some way in the database. For example a stand may be cropped, and consequently much in the database will need to change. Secondly the operational analyst will examine what operations on the data are required. For example, if it is necessary to know the volume of wood in a stand, what processes are necessary to find this item of 'derived data'? In this example (Figure 1), the 'stand-type' (I have assumed stands are a monoculture) and 'stand-age' have to be 'looked up' in the STAND table and cross referred to 'Wood-Type', 'Wood-Age' in the WOOD table and the 'Average-Volume' then multiplied by 'Number of Trees' in the STAND table to produce the result. Clearly, once the stand has been cropped the 'Average-Volume' figure can be changed to the actual value found.

It is important to realise that data analysis is a complex procedure although it *can* be done by almost anyone and seems simple (Veryard, 1984). To get the most benefit out of the final database, data analysis should be done by someone with considerable experience, in conjunction with people having considerable knowledge of the application. Because it is done with pencil and paper and 'hot air', it is cheap and if done well, saves time and considerable amounts of money.

### APPROPRIATE TECHNOLOGIES

Selection of the appropriate DBMS (DataBase Management Systems) to suit the data design and operational analysis is a complex activity full of compromises. Often the decision is not made by a close examination of the data design mentioned above: rather, a computer is bought (for whatever reason) and a database management system and the database mounted, willy-nilly, upon it.

Two courses of action are open to you if you are not to suffer the near fatal but little known disease 'flavour-of-the-month-itis' in the choice of a suitable DBMS. The first is to read all about 'hierarchical', 'network' and 'relational' systems in one or two well chosen books, well in advance (eg Martin, 1975; Ullman, 1981; Date, 1981) and the second is to hire your second expert. He is given the detailed data design and operational analysis and is asked,

a) to pick it to pieces, and
b) to suggest how it could be implemented and for what costs and implementation timescale.

The decision now consists of believing your expert and/or getting a second opinion and/or planning your loans. You should also sneak a look at the two sheets of paper in your valuables safe . . .

Once all this is done order your equipment and watch it being installed. Get your supplier to run through the software, demonstrating everything you can think of, then pay him, arrange the hardware and software maintenance contract and do a little praying.

### DATABASE IMPLEMENTATION

The actual implementation of the database on the DBMS should be done 'prototypically'—that is 'quickly and dirtily' getting as much of your design and operation analysis working as soon as is possible.

Typically, give yourself only between a week (for a small database) and six months *at most* to get something 'up and running'. The purpose of doing this is five fold,

1) otherwise you will be wanting to change the data design or operational analysis;
2) an unused system wastes time and money;
3) you will begin to see all the minor problems quickly;
4) running two systems, one manual and one automatic, is hell to keep up with;
5) you will begin to assess how the system will ‘fit in’.

The last point is perhaps the most important. The people most involved should be using the data in the database, not playing with the new computer system. If they see the new toy in terms of how it helps them do their job effectively, they will think in terms of how to play with the data rather than how to play with the system—a very important psychological point.

Once the prototypical database implementation is running on the chosen DBMS, those inefficient or ineffective bits of the implementation should be re-written as soon as possible. Aim to replace the majority (90 percent) of the prototypical database implementation within the first 18 months. At the end of the 18 months you will also be in a position to review the choice of DBMS.

Once the system is running properly, you will need people to input data, update incorrect data and most importantly to manage the system as a whole. It is important to dream up ways to try and stop the people becoming too priestly . . .

### EXPLOITING THE DATABASE, TUNING AND ENHANCEMENTS

Everyone should be able to use the database. This includes secretaries, managers, everyone. Where necessary, this may mean modifying the system, or expanding it with more access points. A good database is a resource, and if not managed properly, used in appropriate ways, and continuously developed it will stagnate and die, or worse, waste a lot of money. Every system will need tuning, that is, made to run faster. Database systems are notoriously susceptible to being tuned. For example doubling memory allocation and using one index (the jargon here is not really relevant), resulted in a four fold improvement in speed in handling a standard query, in one example with which I was involved. However, if that query was asked only once a month, the total cost in terms of effort to implement it would not have been worth it. Always have a priority list of things which need to go faster and work on them on the basis of overall cost effectiveness.

When tuning can no longer be done to improve matters, an enhancement to the system may become necessary either because you cannot do something or you cannot do something fast enough, despite tuning. The problem with enhancements is that usually the enhancement is more complex than the original system. It may well be that what you really need is a second look at the data design and operational analysis and a totally new system. In the long run this may prove more cost effective than your enhancement! The inevitable conclusion is that you must not let your priests run ahead and implement enhancements. Cost it out, once you have some facts, then let other factors come into play. And finally, be wary of enhancements ‘that are going to save you lots of money’—they may do, but usually don’t.

### EVALUATION PROCEDURES

Your system has now been working for three years, of which the last year has been blissful. Most of the various parts of it are running at the sorts of speeds and effectivenesses you envisaged. In fact all is going well—or as well as can be expected.

You suddenly remember the two sheets of paper in your valuables safe.

The two sheets of paper will remind you what your objectives once were. Comparing this with your achievements will be most instructive. Having done this, (and in passing, assessed your own ability at achieving objectives) be prepared to throw out the current system completely if any of the objectives you read are not met. If some are met, but only somewhat, or some have changed for whatever reason, re-evaluate them—not with the light of hindsight, but with the childish enthusiasm you used to frame the first set. Then break the news to your high priests.

The evaluation of a system, beyond the secret one done in the strong room, should be continuous and involve all concerned. In general, every database needs to be evaluated by the 'triumvirate' of 'users', 'operators' and 'developers'. Do not *always* listen to just the users, occasionally let the operators and developers have a word. It will usually be the case that the users will be happy with what they've got and be happy with a constant stream of cosmetic, minor changes, while it will be the operators and developers that want the major changes. This effect is one which will need careful management, and as ever let cost be the main arbiter.

### FINAL THOUGHTS—APPLICATION TO FORESTRY

If you have read this far you will have noted a certain cynicism, hopefully scented with reality. Let me say I believe that databases *are* useful and when used properly they are of inestimable value to almost all organisations. What they are not is a panacea for all the organisation's problems.

Returning to the major theme of the conference, I shall briefly outline the areas where I feel databases could be of use to forestry. These are:

a) man-management. Not only in all aspects of planning, but sales, salaries and all the other paraphenalia of a large and complex industry;

b) in forest management. Covering yields, forest distribution, the logistics of cropping and removal and all the natural environment aspects;

c) in data management. That is other people's data which are of interest to forestry eg. weather, yields, prices, growth statistics and so on;

d) bibliography. This may be just at the access level; for example you may wish to know what research has been done on the pathology of frost damage;

e) in the production of maps, papers etc for publication. The database may be indispensible here, for example, in area calculations and the presentation of data to statistical/graphical computer programs.

Within the Natural Environmental Research Council, databases in each of these areas are already in existence. They may be called upon by people within NERC and of course we act as a service to outside organisations as well. The fact that we have them and need them is a pointer, to me at least, that there are likely to be similar needs in the forestry business.

### DISCUSSION

I have tried to show that data are a resource in no wit different to a forest. They need management, if they are to be of value. This management must be effective at all stages; from the original planning stage, to the final data archival stage.

The most critical step in the process is the choice of DBMS. Simple single user systems based on perhaps one or two files, while not satisfying the criteria of the

DBMS, do give considerable experience of the sort of thing modern DBMS technology can do. The secret here is not to assume that a limitation on the simple system is a limitation on the technology as a whole. Ultimately, the choice of a suitable DBMS and technology is often a matter of taste, and it may well be governed by many different factors requiring much compromise in the data analysis. Where possible, when such compromises are required—resist them.

The need to implement a database quickly cannot be overstressed. It is almost certain that you will have to replace most of the initial system anyway, so planning for this replacement should be easy and involve little cost penalty. Furthermore, a refinement of the requirement will have occurred and allowing for the incorporation of this into the overall scheme of things is a wise precaution. If, in addition, the system has been made available to every employee, then it will be easy to incorporate their modifications to the data analysis at the same time.

Much of the need for database tuning and the implementation of enhancements stems from an evaluation process which ensures that the database is an animate affair, liable and receptive to change. However, from time to time it will also be necessary to take stock, to sit back and to investigate the original objectives in implementing the database system. Do try and give yourself enough time, free from other pressure, to do this effectively.

Finally, it is my observation that databases if used properly can profitably be used in all aspects of forestry. However, these advantages will come about only by ruthless evaluation of the costs involved in implementing the system and assessing the real motives and objectives in having the system in the first place. If all these caveats are observed, designing and implementing a database can be very rewarding and most of all, considerable fun!

REFERENCES

Date, C.J. (1981) *An Introduction to Database Systems*. 3rd Ed. Allison-Wesley Publishing Co., Reading, Massachusets.

Martin, J. (1975) *Computer Database Organisation*. Prentice-Hall International, Englewood Cliffs, New Jersey.

Townsend, R. (1970) *Up the Organisation*. Baronet Books Hodder Fawcett Ltd, London.

Ullman, J.D. (1980) *Principals of Database Systems* Computer Science Press, Rockville, Maryland.

Veryard, R. (1984) *Pragmatic Data Analysis*. Blackwell Scientific Publications, Oxford, England.

Vetter, M. and Maddison, R.N. (1981) *Database Design Methodology*. Prentice-Hall International, Englewood Cliffs, New Jersey.

# THE DESIGN AND USE OF A DATABASE
# A WORKED EXAMPLE

BERNARD M. DIAZ AND ROBERT H. SANDERSON

*Natural Environment Research Council,*
*Holbrook House, Station Road, Swindon SN1 1DE*

## INTRODUCTION

The previous paper gives some general advice on how to go about the implementation of a database. In this paper, the two main aspects of such an implementation, data design and operational analysis, are illustrated with a worked example. It is based on hypothetical data, but serves to highlight a number of the issues involved. The illustration uses the 'relational model' which presents data in a tabular form.

## SUMMARISING THE CONTENT OF A DATABASE

The previous paper stressed the importance of having the contents of any database listed in that database. The table below summarises the contents of an 'ESTATE' database which is used throughout the example. It may be seen that there are eight tables, the first is ASSESSMENTS, the next COMPARTMENT and so on. Each table, in its turn, has a number of columns corresponding to the attributes that were recorded. For example ASSESSMENTS has three columns named 'a measure', 'a type' and 'o area'. It should be noted that the table itself has three columns, with column headings 'table-name', 'column-name' and 'len' and the name of this table is TABLES.

TABLES

| table-name | column-name | len |
|---|---|---|
| assessments | a measure | 2 |
| assessments | a type | 20 |
| assessments | o number | 2 |
| compartment | c area | 2 |
| compartment | c number | 1 |
| compartment | in forest | 1 |
| forest | code | 1 |
| forest | forest name | 30 |
| forest | map details | 4 |
| operation | c number | 1 |
| operation | day | 2 |
| operation | in forest | 1 |
| operation | o code | 20 |
| operation | o number | 2 |
| operation | s number | 1 |
| operation | year | 2 |
| species | species code | 1 |
| species | species name | 20 |
| stand | c number | 1 |
| stand | in forest | 1 |
| stand | s area | 2 |
| stand | s number | 1 |
| stand | species code | 1 |

| table-name | column-name | | len |
|---|---|---|---|
| timber | age | | 2 |
| timber | authority | | 20 |
| timber | species | code | 1 |
| timber | weight | | 2 |
| wood | code | | 1 |
| wood | species | code | 10 |
| wood | wt@10 | | 2 |
| wood | wt@100 | | 2 |
| wood | wt@120 | | 2 |
| wood | wt@150 | | 2 |
| wood | wt@20 | | 2 |
| wood | wt@300 | | 2 |
| wood | wt@80 | | 2 |

## EXTRACTING DATA FROM THE DATABASE

In order to extract information from the database, we need a language to express our requests. Such languages have a similar structure and the particular one we will use in the examples is called IDL (Intelligent Database Language). The following query, written in IDL, is 'retrieve all the rows of the table COMPARTMENT(abbreviated to c.) and print out the column "compartment area" (c.c_area) where the column "compartment number" (c.c_number) is '1' and where the column "in forest" contains the code "G" (indicating the Grand forest)'.

```
retrieve (c.c_area) where c.c_number=1 and c. in_forest='G'
```

| c_area |
|---|
| 320 |

The result of this enquiry is 320 and there is only one row in the result table because, as one might expect, only one such compartment exists.

## GOOD AND BAD DATA DESIGN

The request above could have been formulated as:

```
retrieve (compart_area=sum (s.s_area where s.c_number=1 and s. in_forest='G'))
```

| compart_area |
|---|
| 290 |

Here, we ask for the 'compartment area' to be calculated as the sum of the areas of the stands which comprise compartment '1' in forest 'G'. Now we get the result 290, which differs from the previous answer.

Which is correct?

Well, I suppose it depends . . .

However, what it does indicate is bad data design. One of the golden rules of data design is to avoid storing data which can be derived. Deletion of the column c_area in the COMPARTMENT table would avoid this particular embarrassment. However, it must now be remembered that deletion of a row in STAND has serious consequences—and this (the fact that the user needs to be aware of the problem) is clearly a better data design since it reflects the real world situation where stands do not just disappear. The design would have been even better if a 'contraint mechanism' existed such that an attempt to change the STAND would warn the user of the consequences.

Let us look at the table WOOD, which contains the following data:

WOOD

| code | species_code | wt@10 | wt@20 | wt@80 | wt@100 | wt@120 | wt@150 | wt@300 |
|---|---|---|---|---|---|---|---|---|
| s | oak | 2 | 10 | 50 | 0 | 120 | 220 | 250 |
| s | pine | 10 | 50 | 110 | 0 | 100 | 90 | 90 |
| s | spruce | 5 | 10 | 100 | 0 | 120 | 100 | 100 |

Note that it has column names which contain useful data, namely 10, 20, 80, and so on. What happens if 'weight at 50 years' for spruce is to be added—well another column must be inserted (wt@50). But what about weight of pine and oak at 50 years? Again—bad design, in this case because it forces us to 'invent' data. The table TIMBER is a better second attempt:

TIMBER

| species_code | age | weight | authority |
|---|---|---|---|
| o | 10 | 2 | average data |
| o | 20 | 10 | average data |
| o | 80 | 50 | average data |
| o | 120 | 120 | average data |
| o | 150 | 220 | average data |
| o | 300 | 250 | average data |
| p | 10 | 10 | average data |
| p | 20 | 50 | average data |
| p | 80 | 110 | average data |
| p | 120 | 100 | average data |
| p | 150 | 90 | average data |
| p | 300 | 90 | average data |
| s | 10 | 5 | average data |
| s | 20 | 10 | average data |
| s | 56 | 60 | Rob's estimate |
| s | 80 | 100 | average data |
| s | 120 | 120 | average data |
| s | 150 | 100 | average data |
| s | 300 | 100 | average data |

PRINTING OUT THE TABLES IN THE DATABASE

Now leaving aside problems of data analyses, we present the remainder of the data in our prototypic implementation. These data are the set we shall use in the queries which follow. They are intended merely to demonstrate the sort of thing that would be possible if a database management system were available. It is worth noting that we have chosen a very powerful system for the implementation—this was to speed up the whole design and implementation process. Many of the facilities we demonstrate will be available on smaller (cheaper) systems although the actual ways of specifying them will be different.

The first table is SPECIES. We list it with the retrieve command. The bit in brackets

(.all) indicates that we want all the columns to be printed out. The 'order by' clause specifies that we want the table sorted into a specific order, in this case it is alphabetical. We will use a similar command to list all the tables.

SPECIES

1) retrieve (species.all) order by species code

| species_code | species_name |
|---|---|
| o | oak |
| p | pine |
| s | spruce |

FOREST

1) retrieve (forest.all) order by code

| Code | forest_name | map_details |
|---|---|---|
| C | Calder | NM23 |
| D | Dean | SO71 |
| G | Grand | NM44 |
| K | Knaresborough | SE40 |
| N | New | SU34 |

COMPARTMENT

1) retrieve (compartment.all) order by in_forest, c_number

| in_forest | c_number | c_area |
|---|---|---|
| C | 1 | 100 |
| C | 2 | 120 |
| C | 3 | 90 |
| C | 4 | 100 |
| C | 5 | 70 |
| G | 1 | 320 |
| G | 4 | 170 |
| G | 5 | 200 |
| G | 6 | 270 |

STAND

1) retrieve (stand.all) order by in_forest, c_number, s_number

| in_forest | c_number | s_number | species_code | s_area |
|---|---|---|---|---|
| G | 1 | 1 | s | 10 |
| G | 1 | 2 | s | 50 |
| G | 1 | 3 | s | 90 |
| G | 1 | 4 | p | 20 |
| G | 1 | 5 | p | 30 |
| G | 1 | 7 | s | 10 |
| G | 1 | 9 | o | 80 |

OPERATION

1) retrieve (operation.all) order by year, day

| o_number | year | day | in_forest | c_number | s_number | o_code |
|---|---|---|---|---|---|---|
| 11 | 1904 | 301 | G | 1 | 1 | planted |
| 212 | 1924 | 301 | G | 1 | 1 | thinned |
| 216 | 1924 | 321 | G | 1 | 1 | assessed |
| 992 | 1960 | 256 | G | 1 | 1 | assessed |
| 994 | 1960 | 258 | G | 1 | 1 | thinned |
| 7000 | 1969 | 147 | G | 1 | 1 | assessed |
| 9413 | 1979 | 310 | G | 1 | 1 | assessed |
| 12014 | 1984 | 302 | G | 1 | 1 | assessed |
| 15012 | 1984 | 329 | G | 1 | 1 | cropped |

ASSESSMENTS

1) retrieve (assessments.all) order by o_number

| o_number | a_type | a_measure |
|---|---|---|
| 216 | pH | 6 |
| 216 | density | 200 |
| 216 | quality | 1 |
| 992 | density | 190 |
| 7000 | pH | 6 |
| 7000 | density | 155 |
| 7000 | quality | 3 |
| 7000 | nutrition K | -5 |
| 9413 | density | 150 |
| 9413 | nutrition K | -7 |
| 12014 | density | 140 |
| 12014 | quality | 2 |
| 15012 | yield | 1020 |
| 15012 | quality | 1 |

ASKING QUESTIONS: SOME EXAMPLE QUERIES

We illustrate what is possible when a database approach is available with two example queries. For simplicity we have again used the IDL 'query language' as our database access method. A retrieval in this language has the following prototype form:

retrieve (what clause) (ordering clause) (where clause)

The 'retrieve' command itself is followed by three clauses. The 'what clause' tells the computer what columns to retrieve, that is the column names which should be in the result table. The 'ordering clause' states how the data in the table should be presented to the user. Finally, the 'where clause' indicates the search predicates which must be satisfied by the query.

Query 1. In this the user wishes to find what operations have been done on stand '1' of compartment '1' in the Grand Forest. This is achieved with the following query:

```
retrieve (o.all) order by o_number where o.s_number=1 and o.c_number=1 and
o.in_forest='G'
```

| o_number | year | day | in_forest | c_number | s_number | o_code |
|---|---|---|---|---|---|---|
| 11 | 1904 | 301 | G | 1 | 1 | planted |
| 212 | 1924 | 301 | G | 1 | 1 | thinned |
| 216 | 1924 | 321 | G | 1 | 1 | assessed |
| 992 | 1960 | 256 | G | 1 | 1 | assessed |
| 994 | 1960 | 258 | G | 1 | 1 | thinned |
| 7000 | 1969 | 147 | G | 1 | 1 | assessed |
| 9413 | 1979 | 310 | G | 1 | 1 | assessed |
| 12014 | 1984 | 302 | G | 1 | 1 | assessed |
| 15012 | 1984 | 329 | G | 1 | 1 | cropped |

As before, we have abbreviated the OPERATION table name to 'o.'. The '.all' as outlined above is a shorthand allowing us to specify that we wish to see the contents of all the OPERATION table columns in the result table. If we examine the OPERATION table we will see that it has seven columns. We can see that the where clause indicates that only the rows where the stand number (s number) is '1', the compartment number (c_number) is '1' and the in_forest code is 'G' are required.

Query 2. A slightly more ambitious query this time to find those assessments made after 1983 and what they achieved.

```
retrieve (o.year,o.day,a.a_type,a.a_measure) order by year,day where
a.o_number=o.o_number and o.s_number=1 and o.c_number=1 and
o.in_forest='G' and o.o_code='assessed' and o.year > 1983
```

| year | day | a_type | a_measure |
|---|---|---|---|
| 1984 | 302 | density | 140 |
| 1984 | 302 | quality | 2 |

If the ESTATE tables are examined, it will become apparent that there is no single table which contains the data required to satisfy this query. In particular you will note that there are no dates in the ASSESSMENT table and no assessment type information in the OPERATION table and consequently a link between the two tables is required if the query is to be satisfied. This is achieved by using the fact that that all the assessment measures made on one day have the same operation number, hence the construction 'a.o number= o.o number'. (Remember that many things could be measured on an assessment outing eg pH, wood quality, wood density, etc.). With this link (or join) in place we can now think of the two tables as one gigantic one and enter the remaining predicates as though this was the case. Thus, we can see that 'o.o code= "assessed" ' and 'o > 1983' etc. It is important to note that two tables are referenced in this query. It is usually the case that the more complicated the query the more tables are required to satisfy it.

## SUMMARY COSTINGS

We made reasonable estimates of the time spent in generating this example database. From these figures, we calculated the cost of implementing a complete system based on the data used in this example. Five man-days were required to design and implement the ESTATE database at a cost of around £500. Most of this time was spent in generating very complex stored or 'canned' queries which avoid the necessity for the user to learn IDL. Although none of these have been presented in this paper, it is fairly certain that the majority of serious database work on ESTATE would be done using this access method. Based on assumptions such as these, we estimate that the cost of designing and implementing a real version of ESTATE for a large estate would involve approximately 50 Mbytes of data and would cost in the region of £20,000—this would include the purchase of hardware. Annual running costs could be £50,000 consisting almost entirely of salaries and associated overheads and this result coincides with the conclusion made in the previous paper.

# THE FORESTRY COMMISSION SUBCOMPARTMENT DATABASE

J. DEWAR

*Forestry Commission, Alice Holt Lodge, Farnham*

SUMMARY

The need in forestry for site and crop data in the form of a database is discussed. A brief description is given of the Forestry Commission's subcompartment database, and various uses of the database are considered.

INTRODUCTION

In any business managers are concerned with, among other things, collecting and manipulating data to give useful information about the resources used in the organisation—money, people, equipment and materials. The information obtained is intended to allow managers to make better decisions and exert better control over these elements and to ensure that they are used to achieve the organisation's objectives as efficiently as possible.

In forestry, there is an additional resource which can be looked upon as a subset of materials but which is sufficiently different and important to justify separate consideration. This is the growing stock and land associated with it. The growing stock is different from other materials in that it changes with time. It is also different in that it is geographically widely dispersed. It is important since, within any forestry enterprise which owns forests, the capital locked up in land and plantations is so large as to dwarf investment in machinery, equipment and stocks of other materials. The annual value increment of the growing stock is a major source of added value arising from the organisation's activities. Managing such a large investment requires sound decisions about the growing stock and land, such as what to plant, where and when to thin or fell and many other related decisions. It follows that poor decisions can be very expensive, and certainly as expensive as poor decisions about the management of other resources. Hence the need for separate data about the forest covering the land and the growing stock.

THE NEED FOR A DATABASE

When collecting data it is generally as well to know the purpose to which the data will be put. The danger however is that a series of *ad hoc* surveys are carried out to answer specific questions. Thus data could be collected which would allow the following questions to be answered:

What is the area of each species in the forest?
What is the age class distribution?
What is the area of each soil type?

but without further information it is not possible to answer the question—

What is the area of Sitka spruce over 40 years of age on surface water gley?

This can only be done if the information about species, age and soil is related to a fourth item, a unique location. It is surprising how often data have been collected during different surveys which, because precise locational information was not collected or is not compatible, can not be used in combination without a lot of additional work. The usual locational address used in forestry is the compartment which may be sub-divided to fit some features of site or crop into subcompartments. Within any forest area sharing the same general location name the subcompartment address will be unique.

Another danger is that information is stored in a computer in such a way that it can only be easily accessed to answer pre-determined questions. Writing a program to answer another question may either be impossible or adversely affect existing programs. The response of users to this sort of problem is to start keeping their own set of records designed for their own purposes. This, however, leads to duplication of effort and, if updating of data is not carried out by all users at the same time, inconsistencies arise. The answer to this problem is to create a database.

A database is defined as a collection of stored data organised in such a way that all user data requirements are satisfied by the database. In general there is only one copy of each item although there may be controlled repetition of some data.

The advantages of a database are that data are independent of the program that use them, data can be easily added to or reorganised and new programs can use the data in a modified form without affecting programs which use the same data.

## THE FORESTRY COMMISSION SUBCOMPARTMENT DATABASE

Within the Forestry Commission subcompartment information was initially recorded on edge punched cards from which summaries of areas by species, age class and yield class were compiled. This information was then used for making production forecasts by laborious manual calculation but subsequently was computerised. It was only in 1976 that the subcompartment data was transferred to computer storage and only recently has it been organised and stored in a form which merits being called a database.

Subcompartment data is held in three fields or groups.

1. The key field provides location by Forest District, Geographical Block, Compartment and Subcompartment. Components of mixed crops are also noted in this group.
2. The required data field contains crop details of species, age, yield class and area, other land use data and administrative information such as Local Authority area.
3. The optional data field contains details of site factors such as soil type and windthrow hazard class, and codes that group subcompartments into common treatment blocks.

For further information on the contents of the database see Horne and Whitlock (1984).

Updating the data to 31 March each year is the responsibility of local managers with Field Survey staff checking and giving advice as necessary. In addition the data are fully revised following a survey every 12 to 15 years. A revised subcompartment print-out, listing the full computer data holding, is produced annually after updating and full revision. This is available to Forest District staff for reference although, for many purposes, being a bulky document it is not as useful as being able to call up information on a Visual Display Unit.

## USES OF THE DATABASE

In considering the uses to which the database can be put I have found it convenient to recognise three major categories. These are:

1. Sorting, listing and summation.
2. Simulation and optimisation.
3. Pattern detection.

Sorting, listing and summation is a simple routine chore which is tedious to do manually but which has long been recognised as a suitable task for computers. By sorting through the data to find each subcompartment with, say, Sitka spruce present and summing the area of each such subcompartment the total area of Sitka spruce can be obtained. Groups of information can also be identified, sorted and summed to answer questions such as: what is the area of Sitka spruce yield class 12 planted in 1947 on Windthrow Hazard Class V sites? Subcompartments with this combination present can also be listed. This is not only of value to local management in identifying crops suitable for thinning or felling but has been used nationally in carrying out surveys of crops most as risk from a particular pest such as *Dendroctonus* where a listing of the location of vulnerable crops is required. It has also been used for stratified random sampling of compartments in different areas with different crop and site characteristics in order to survey crop health.

Simulation is a technique whereby actions are applied to a model of reality rather than reality itself to find the effect of different actions. The major simulation involving the subcompartment database is production forecasting. Yield models have been produced for all major combinations of species, yield class, spacing and type of thinning which show the effect in terms of volume production per hectare of applying different cutting regimes (ie. timing and type of thinning and age of felling) to different types of crop. The growing stock database can be looked on as a model of the forest and a production forecast is obtained by making assumptions about the cutting regimes to be used, applying the appropriate yield models to each subcompartment and summing the volumes. The assumptions about cutting regimes can of course be changed so that the forest manager can test the effect in terms of volume production of following different courses of action.

The need for accurate forecasts of future timber production cannot be over-emphasised. Timber-using industries will only carry out major investments if they have some degree of assurance of future supplies. Because timber-growing is such a long-term business, supplies for the next 10 to 20 years are determined by crops already existing. Changes in supply can only be achieved by altering the cutting regimes applied to existing crops. If there should be a shortfall in supply, then any alterations to cutting regimes to achieve this extra volume will almost certainly cause reductions in volume production at a later date. Thus what may start off as a small shortfall can, if corrected by overcutting, lead to an even larger shortfall at a later date. It is therefore important not only that accurate forecasts of production are made but also that managers consider the effect in terms of future production of adopting different cutting regimes. This the production forecasting system using the subcompartment database allows them to do.

Optimisation of forest management practices is done regularly without reference to the database. Examples are calculations of optimum felling ages and calculations of optimum road network densities. In many circumstances there is no need to refer to the database because generalised solutions can be developed which would apply to all crops with the same characteristics. Thus if the optimum felling age for Sitka spruce of yield class 12 on stable sites is age 50 this would apply to all compartments with these characteristics. However what may be optimal for one subcompartment viewed in isolation will not necessarily be optimal for all such compartments if there are other factors to be taken into account besides maximising Net Discount Revenue (NDR), in other words if there are constraints to be satisfied. These may take the form of landscape considerations, resource limitations or marketing constraints. Management needs to try to reduce the effect of such constraints but it may not always be possible to eliminate them entirely. A solution may be optimal in terms of maximising NDR but unacceptable in terms of satisfying markets, particularly where there is a contractual obligation.

Optimisation of cutting regimes to maximise NDR and satisfy a constraint has been done crudely using simulation techniques to test the effect of varying optimal cutting regimes with the cost of meeting the constraint being calculated as a second stage. There is however a well tried technique in the form of linear programming for doing this sort of work. In theory, it would be possible to add financial data to the subcompartment database and, using linear programming, calculate the solution which maximises NDR while meeting one or more constraints. However, to do this requires information on future costs and income for each crop at various stages of development and sufficient computing power to handle the quite massive calculations involved. Therefore, it may be that the cost of attempting this sort of analysis on a large scale may currently be greater than any benefits obtained. However, I have little doubt that in the longer term this will become an increasingly important application of the database.

Pattern detection is the third major category I wish to discuss. It is part of the art of management to recognise when an event is not random but is linked to some other factor which may be in the manager's control. An example is the recognition that the volume of timber produced as a result of planting Sitka spruce on a particular site is generally greater than if larch had been planted. The difference is large enough to have been apparent to foresters without the aid of computers but analysis of the database confirms that on similar sites Sitka spruce has a higher yield class than larch. Other patterns may not be so obvious but their recognition could still lead to significant benefits. A more recent example is the recognition of the relationship between height of onset of windthrow and the various site factors which became the basis of the windthrow hazard classification. It may well be that there are many other factors of management significance which are correlated with features of site or treatment awaiting detection. Such patterns can be detected without computers using feature or coincidence cards but transfer of the data to the cards is a slow business. However it ought to be possible to display information from the database on a Visual Display Unit which would achieve the same result. Coincidence of factors does not of course guarantee a causal connection but it could serve to point researchers in the direction of fruitful areas of investigation.

## CONCLUSION

The Forestry Commission Subcompartment Database has grown out of the data collected primarily for answering simple questions on the nature of the growing stock and for production forecasting. The production forecasting system in particular is tried, tested and successful in that it gives reliable results quickly and more cheaply than any non-computerised system. With this foundation we have been able to add to the total stock of data, form a database and consider applications. Experience suggests that this is the right approach rather than to start from scratch with a grand scheme intended to answer all conceivable questions. The result of the latter approach is likely to be disappointing.

REFERENCES

Horn, A.I.D. and Whitlock, M.D. (1984) The Forestry Commission Subcompartment Database I. Description of the Data Held. *Forestry Commission Research Information Note,* 94/84/FS.

# STAND GROWTH MODELLING—THE VALUE TO FOREST MANAGERS

K. RENNOLLS AND T.J.D. ROLLINSON

*Forestry Commission Research Station, Alice Holt Lodge, Farnham, Surrey, UK.*

SUMMARY

Some of the limitations of traditional yield tables are discussed and the advantages of using computer-based simulation growth models are presented. The model currently under development by the Forestry Commission, which depends on individual trees competing for limited resources, is briefly described and set within the context of other simulation models of forest growth. Possibilities for the long term widespread use of computer models in forest management are highlighted.

## INTRODUCTION

Traditional yield tables have been used as an aid to forest management in Europe for over 150 years (Osmaston, 1968). In Britain the main use of yield tables is to compare alternative management treatments and to forecast production (Edwards and Christie, 1981). The principal disadvantage of yield tables is that they provide descriptions of growth for that narrow range of management regimes for which growth and yield plots exist. In addition, they are usually developed at a stand level, and, while providing adequate information about total yield production, give little detail about how this is distributed between individual trees which is necessary to forecast the products which can be cut from a stand.

If forest managers are to make informed decisions about which management treatments to use they must have information which allows them to compare and select the complete range of possible treatments. Computer-based simulation models of tree growth have the potential to predict crop performance under a wide range of management treatments (Sutton, 1982). This paper considers some of the advantages of such models, and describes some current developments with particular reference to a model being developed by the Forestry Commission Research Division at Alice Holt. In the final section, future developments and uses of simulation models by forest managers are discussed.

## COMPUTER BASED GROWTH MODELS

A model is an abstract representation of some aspect or system in the real world and as such can form the basis for our understanding and deductions about the real world. The descriptive use of a straight line to represent a scatter diagram of points on a piece of graph paper is a very simple example of a model. A complex model, on the other hand, may include mathematical description of the biochemistry underlying photosynthesis and nutrition.

A model is called a 'computer-based simulation model' when its structure is programmed into a computer which is then used to calculate the dynamic behaviour of the system being modelled. The hope is that this will parallel the dynamics of the corresponding aspect of the real world.

Both Thornley (1980) and Grossman (1983) stress that there is a hierarchy of levels at which consistent and integrated models can be developed. Grossman emphasises that modelling can only be predictively successful if the modelling techniques are appropriate to the level of the real world at which questions are asked. Before describing the model currently being developed within the Forestry Commission Research Division we review some of the models that have been developed so far and consider if they are at the appropriate conceptual level to be of use to forest managers.

*Stand models*

Considerable effort has been expended in the past two decades to develop stand-level models (ie not involving individual trees) which describe the development of diameter, height and volume distributions in a consistent way, (see Clutter *et.al.* 1983 for a general review of growth models and Munro (1984), Pollanschutz (1983), Trimble and Shriner (1981) and Vanclay (1984) for indications of the current state of the art). Such models include and extend conventional yield table methods by expressing them in mathematical form. They are of great practical utility since their variables are those most well known to foresters (dbh, top height etc) and when implemented on a computer are efficient in computer time.

*Individual tree models*

More detailed models of stand growth use the development of an individual tree, in competition with its neighbours, as the basis of the model; stand growth is obtained by summing the growth of individual trees.

Distance-independent individual tree models do not require information about the spatial distribution of trees, but make use of a crown-competition factor for each tree. These are economical in computer time and data capture requirements and have been applied in the USA (see Stage (1973) and Wykoff (1983)).

Newnham (1964), Mitchell (1975) and Ek and Monserud (1974) have developed models in which it is necessary to know the distances to each tree's neighbours. These distance-dependent models include the normal mensurational variables and allometric relationships and provide the forester with the ability to evaluate different thinning and spacing treatments. However they do require greater computer storage, are relatively time consuming to run and have, to date, been little used in practice. With the increasing power and decreasing cost of computers it may be feasible to make much wider practical use of these.

*Biologically based models*

It is questionable whether empirical growth models, such as those mentioned above, are of much value when attempting to answer questions concerning, for example, the growth response of crops to new combinations of thinning and fertilizer treatment, or the reduced growth induced by defoliation or drought. In recent years there has been a move to develop biologically-based models which include within their structure those underlying biological factors and processes which control and determine tree and forest growth. This underlying structure will increase our confidence when we attempt to use the models to predict in essentially new circumstances. However, we must ask: 'how much biological detail should be included in models designed to aid and improve forest management practices?'. Thornley (1980) argues that too much emphasis on fundamental research can be counterproductive with respect to the stated objective. There is a strong case for the most appropriate level of modelling being that which is a hybrid between the traditional empirical forestry yield modelling approach and that which takes a holistic ecological approach to the forest crop, (see also Kimmins and Scoullar (1983)). Such models provide likely avenues to a major advance in stand growth modelling in the short term.

*Physical/physiological models*

Finally we should mention the research done in developing 'physical/physiological' models of the forest ecosystem. Such models are those that apply the laws of physics to the process of matter, energy and momentum transfer. The physiological side to this approach, which attempts to model the fundamentals, will characterise the basic biological aspects of photosynthesis and partitioning of assimilates. This level of approach was that adopted by the International Biological Program, and by the SWECON project, both of the 1970's. Although there have been many valuable research results from these projects, there is a fairly widely held view that these techniques represent a level of reductionism (see Thornley 1980) which is not appropriate (Grossman 1983) to the questions that have to be answered in forest management.

Models which combine traditional mensuration variables with ecological and physiological submodels, at an appropriate level, are likely to provide the longer term advances in forest growth modelling. See Jarvis (1981), Landsberg (1981) and Rook *et.al.* (1984) for insights as to the level of physiology that is appropriate to forest growth modelling.

## AN INTEGRATED FOREST PROCESS MODEL (IFPM)

A growth model being developed by the Forestry Commission attempts to combine simplified biological processes with an individual tree growth model (Rennolls, 1983). Its structure is shown in Figure 1. The main objective of this model is to relate management actions, such as thinning and fertilisation, to the growth and mortality of a forest stand. This link is established by a distance-dependent individual-tree growth model. Each 'tree' has an internal structure which includes elements of its age-distributed canopy, stem, roots and the associated litter layers. The straight edged boxes in Figure 1 represent the component parts of the model each of which has a number of associated state variables. The flow of resources in the system is indicated by dashed arrowed lines, the rates and directions of these flows being influenced by the competition and litter-fall processes, and the management action of fertilisation. These processes therefore determine the rate of change of some of the state variables of the system. Spacing and thinning are also regarded as processes in the system since they directly influence the state variables associated with the 'stand structure' of the system. The functional dependency of stand structure upon the management process is indicated by a full arrowed line in Figure 1, as is the functional dependency of the growth and mortality processes upon the states of the individual trees in the model. The growth and mortality processes feed back to influence the stand structure of the model so that the model may be regarded as a dynamic control-system in which the controllable input variables are the management actions.

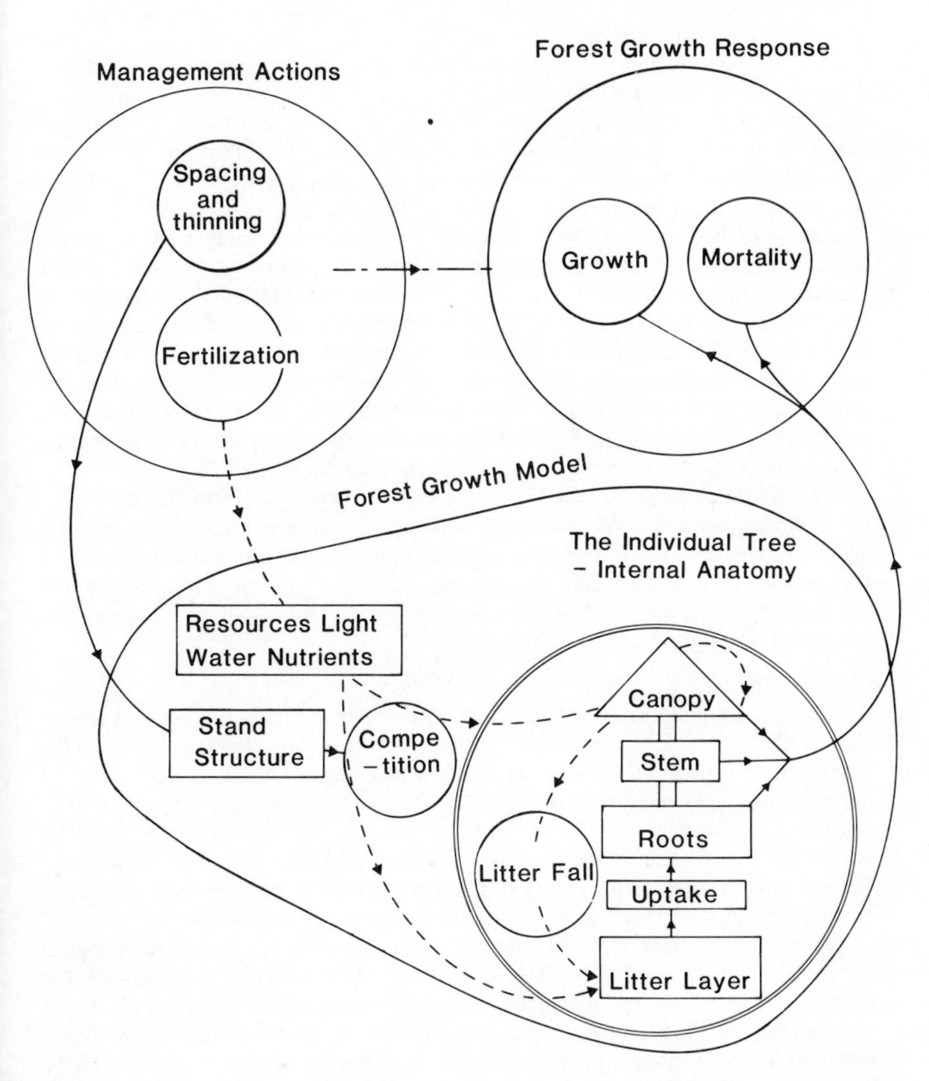


Figure 1. A system diagram of the Integrated Forest Process Model (IFPM). Straight-edged boxes represent the 'component objects' of the model whilst the curved edge boxes represent processes or operations on or between these 'objects'.

This model has been fitted to some of the data from the Forestry Commission sample plot data base, the parameters of the competition, growth and mortality processes being those estimated. The requirement of the model for information about the location of each tree in a stand has resulted in an extended programme of data collection from sample plots.

The model parameters estimated from sample plots subjected to a high intensity of thinning are found to predict well the growth trends of sample plots in which only dead trees are removed. Nine parameters within the competition, growth and mortality processes seem to be capable of representing the full range of thinning intensities. The next stage in the validation of the model is to examine its predictive performance for different spacings and for line thinning. Some preliminary study has been made of the predicted effects of partial defoliation. For example, a fifty percent defoliation of our calibration plots at age 24 is predicted to loose 6.6 years of diameter growth over the subsequent 15 years and the growth rate at the end of this period to be only 61% of what it would have been without defoliation. This prediction is based on the assumption of a complete loss to the system of the nutrients in the defoliated leaves and we would certainly not put too much confidence in such predictions at this stage. We may also use the model to examine the effects on growth of increases in the rate of litter fall and decreases in the rates of internal cycling and nutrient uptake, effects which may well arise as a result of acid deposition.

It is envisaged that this model, or its development, will form the core of a management tool which will be used for a number of purposes. Firstly, the model will allow examinations of the growth and interactions between a wide variety of silvicultural treatments, such as spacing, thinning, pruning and fertilisation. Secondly, the model may be used as a flexible forecasting tool in optimising the management of large forest areas.

## DISCUSSION

'To date . . . computers have mainly been used to process data and not as a tool to expand decision making capabilities' (Keetch, 1982). Jeffers (1964) emphasises the importance of 'decision' models as those which may be used to determine the optimum strategy from amongst a range of alternatives. However, he also says 'The gap between what is possible by modern (operational) research techniques and what is actually being done in the field of forestry is not only large but increasingly exponentially'. Though this gap may not have increased exponentially in the last twenty years it is certainly at least as large today as it was in 1964.

Many authors have stressed that there is a hierarchy of levels of biological integration. Corresponding to these levels there will eventually be a hierarchy of models ranging from those appropriate to the management of a single forest compartment to those appropriate to the management of the whole of the national forest sector (see Grossman (1983)). Models at higher levels of the hierarchy will be integrated combinations of many submodels. The time is not too distant when it will be feasible to link together models of deer and insect population growth and spread (and their damaging effects), national forest data-bases, growth and risk (eg windthrow) models to investigate a wide range of future scenarios. Of course, these goals are easier to state than achieve. Great effort and investment is required in forest growth modelling if the potential benefits are to be realised.

The principal advantages to forest management of computer-based growth models are likely to be:

1. The complexity of model that is possible on a computer goes far beyond what is feasible by purely graphical or mathematical methods. A wealth of information can be concentrated into one easily accessible source, and interrelationships between different parts of the system may be represented.
2. Output from a model can readily be input to an investment appraisal package or other applications using computers.
3. A well constructed simulation model which predicts well for management practices for which data are available, might be used with some confidence to make predictions for practices for which relatively few data are available. Thus it may be possible to limit expensive field trials in testing the complete range of silvicultural treatments.
4. A validated model can generate a conventional yield table for a specific management treatment or combination of treatments quickly and cheaply.
5. Simulation models serve as a valuable aid to research as they can be used to indicate where there are gaps in existing knowledge, and to help to order research priorities.
6. Lastly, a simulation model will be a useful tool for forestry education and training.

Powerful as they are, models are only a tool and hence can only be as good as the assumptions built into them. To satisfactorily specify the initial conditions for the running of forest growth models the exactitude, experience and understanding of forest managers will be indispensable, and of course they, together with the rest of society, are responsible for determining the criteria used to judge the predictions of any model.

To be of practical use, models will have to be easy to use ('user friendly') to allow immediate access by people with little or no computer training and expert systems will feature prominently. As the cost of computing power falls at the same time as the power of computers rises, an increasing number of forest managers are gaining access to computers. As they gain experience and begin to understand how a computer can be used as a tool of forest management, the demand for simulation models and management packages will increase.

ACKNOWLEDGEMENTS

Helen Redshaw and Paul Blackwell have provided much programming assistance in the development of IFPM.

REFERENCES

Clutter, J.L.; Fortson, J.D.; Pienaar, L.V.; Brister, G.H. and Bailey, R.L. (1983) *Timber Management: A quantitative approach.* John Wiley and Sons.

Edwards, P.N. and Christie, J.M. (1981) Yield Models for Forest Management. *Forestry Commission Booklet No. 48,* HMSO.

Ek, A.R. and Monserud, R.A. (1974) FOREST: A computer model for simulating the growth and reproduction of mixed species forest stands. *Research Report A2635, School of Natural Resources.* University of Minnesota.

Grossman, W.D. (1983) *Systems approaches towards complex systems.* In Messerli, P. and Stucki, E. (Eds.): Colloque International MAB-6: Les Alpes Modele et Synthese. Pays-d' Enhaut, 1-3 Juin. Fachbeitr. Schweiz. MAB-Information Nr. 19, Bern.

Jarvis, P.G. (1981) Plant water relations in models of tree growth. *Studia Forestalia Suecicia* **160**, 51-60.

Jeffers, J.N.R. (1964) Mathematical Models in Forestry Research. *Comm. For. Rev.* **43**, 159-168.

Keetch, M.R. (1982) Micro-computers in Forestry Planning. 138-140 in *Microcomputers; A Tool for Foresters;* J.W. Moser (ed.). Purdue University, West Lafayette, Indiana.

Kimmins, J.P. and Scoullar, K.A. (1983) *Forcyte-10, A user's manual.* Faculty of Forestry, University of British Columbia.

Landsberg, J.J. (1981) The number and quality of the driving variables needed to model tree growth. *Studia Forestalia Suecicia* **160**, 43-50.

Mitchell K.J. (1975) Dynamics and simulated yield of Douglas-fir. *Forest Science Monograph* **17**.

Munro, D.D. (1984) Growth Modelling for Fast-Growing Plantations: A Review. IUFRO proceedings; *Symposium on Site and Productivity of Fast Growing Plantations.* Pretoria and Pietermaritzburg, South Africa, 333-334.

Newnham, R.M. (1964) *The development of a stand model for Douglas Fir.* Univ. of British Columbia Fac. of For., unpublished. PhD thesis.

Osmaston, F.C. (1968) *The Management of Forests.* Allen and Unwin, London.

Pollanschutz, J. (Ed.) (1983) Forest Growth Modelling and Simulation. IUFRO Proceedings, Vienna. *Mitteilungen der Forstlichen Bunderversuchsanstalt,* Vol. 147.

Rennolls, K. (1983) The Prospects of an Integrated Forest Process Model, 159-171. In Pollanschutz,, J. (Ed.), Forest Growth Modelling and Simulation. *Mitteilungen der Forstlichen Bundesversuchsanstalt.* Vol. 147.

Rook, D.A.; Grace, J.C.; Beets, P.N.; Whitehead, D.; Santantonio, D. and Madgwick, H.A.I. (1984) In IUFRO Proceedings, *Attributes of trees as crop plants.* In press.

Stage, A.R. (1973) Prognosis model for stand development. *USDA For. Serv. Res. Paper INT-137,* Intermountain For. and Range Exp. Stn, Ogden, Utah.

Sutton, W.R.J. (1982) *A Report from the New Zealand Radiata Pine Task Force.* IUFRO Circular P2.02.02. Newsletter No. 3.

Thornley, J.H.M. (1980) Research strategy in the plant sciences. *Plant, Cell and Environment,* 3, 233-236.

Trimble, J.L. and Shriner, C.R. (1981) *Inventory of United States Forest Growth Models.* Oak Ridge National Laboratory, Oak Ridge, Tennessee 37830.

Vanclay, J.K. (1983) *Techniques for Modelling Timber Yield from Indigenous Forests with special references to Queensland.* Unpublished. MSc thesis. Oxford University.

Wykoff, W.R. (1983) Predicting basal area increment for individual northern Rocky Mountain conifers. 127-143. In Pollanschutz, J. (Ed.). Forest Growth Modelling and Simulation. IUFRO Proceedings Vienna. *Mitteilungen der Forstlichen Bundesversuchsanstalt.* Vol. 147.

# ECONOMIC MODELLING—A WAY TO IMPROVE DECISION MAKING

G.R. WATT

*Forest Economist, Economic Forestry Group, 27 Rutland Square, Edinburgh*

SUMMARY
In this paper economic modelling is examined conceptually and in relation to different categories of models that can be developed. Advances in modelling resulting from the use of computers are then examined. The reasons for and benefits arising from the construction of economic models are discussed, and some examples are given of the principal types of models that have already been developed ranging from global models to detailed financial ones for individual stands of trees. Attention is drawn to the importance of validating economic models prior to their use. It is concluded that computers have significantly expanded the scope and extent of economic modelling, and while not a complete solution, they have considerable potential for further assisting both managers and investors with decision making in forestry.

## INTRODUCTION

The aim of this paper is to consider what is meant by economic modelling, the purpose of building economic models, the different types of models that have already been developed, and how they can help in the forest sector. This broad approach to the subject has been adopted for two reasons. Firstly, a detailed discussion of a particular economic model is likely to be of a limited interest. Secondly, there has been a considerable amount of work done in economic modelling both inside and outside the forest sector, in this country and overseas, and it is important to keep in touch with this work.

## WHAT IS ECONOMIC MODELLING?

Most economic theories are based on a model of one form or another and many economic situations, and the consequences which flow from them, can be expressed mathematically. The use of computers has allowed more complex models to be developed because computers have exceptional capabilities for performing arithmetical operations quickly and cheaply, and for solving well defined economic problems that can be attacked by a rigorously logical procedure (Navon, 1973).

Economic models can very broadly be divided into two categories (Wardle, 1971):

i) *Simulation Models* which are designed to show the financial or economic consequences of possible courses of action to the investor, adviser or manager. In other words they provide a way of examining the overall behaviour of a system.
ii) *Optimising Models* which select courses of action according to objectives, requirements and restrictions. The types of models discussed in this paper are based on either linear programming, or input-output analysis.

WHY DEVELOP ECONOMIC MODELS?

A major problem in making decisions is the difficulty of predicting the outcome of a particular course of action. This is because the human brain is very poor at deducing the behaviour of a complex economic system with many interacting factors (eg Loasby, 1976). However, this is a situation to which computers are well suited, since, provided a valid economic model of a particular situation exists, the consequences of taking different decisions can be examined in advance quickly and in great detail. Economic models therefore have enormous potential as a tool for improving decision taking.

SOME APPLICATIONS OF ECONOMIC MODELS OUTSIDE THE FOREST SECTOR

The applications for which economic models have been developed are many and varied but the following would appear to be the most relevant to the forestry sector.

*Global Models*

One of the first and probably best known global models was that developed on behalf of the Club of Rome (1972) who based their report called 'The Limits to Growth' on its results. It argued on the basis of an elaborate computer simulation model that finite physical resources and pollution would place unavoidable barriers on the progress of world population and economic growth, and that population would not naturally stabilise at the ultimately sustainable level, but would overshoot and then collapse.

Although this model is sometimes regarded as an economic one, it is probably more appropriate to classify it as a physical one. The reason for this is that it took no account of the influence of the price mechanism and the consequent effect that changes in price would have, for example, on substitution, technological change and diversification.

Similar shortcomings exist in the model on which the U.S. Government's Global 2000 report was based (U.S. Government, 1982), but it is interesting to note that the report specifically identified the rapid decline of the world's forests as one of the most serious ecological problems. However, in spite of their limitations, these two examples illustrate the enormous potential of computer based global economic models, if properly constructed, for examining the consequences of taking different decisions in relation to global resources as well as economic and social policies.

*National Economic Models*

In recent years an increasing number of simulation models of national economies have been developed for short term forecasting of such items as economic growth, inflation, unemployment, balance of payments and average earnings. The largest model in this country and probably the most important is the one developed by the Treasury. This has some 700 economic relationships determining the main economic variables, many of which interact with each other. For example, personal consumer spending is related to personal incomes after tax, prices, interest rates and financial wealth. The model also contains detailed treatments of the tax, public expenditure and monetary sectors because of the Treasury's policy interests (Treasury, 1981).

Although forecasts from such models can often show quite wide variation resulting from differing assumptions, they can have relevant implications for the forestry sector. For example, if economic growth and house building are predicted to increase significantly in a year's time, it may be better to delay making a timber sale until then in expectation of receiving higher prices. If on the other hand economic growth is forecast to slow down, it may be judged advisable to sell the timber immediately because prices are likely to fall in the short term.

Another type of economic model is the input-output model which is an attempt to quantify the structure of the economy in detail and to examine the relationship between different economic sectors. For example, an input-output model might show the total value of machines and equipment purchased by the forest sector. From this it is possible to quantify how economically dependent the forest sector is on the manufacturing sector and therefore what impact expansion or contraction of either sector might have on the other. These models have relatively limited value for the forest sector in this country at present, but in others where the sector is relatively large they have been used for examining such issues as employment, foreign trade and tariffs.

## APPLICATIONS WITHIN THE FOREST SECTOR

### *Sector Models*

There has been only one published account of a comprehensive model of the United Kingdom forest product sector as a whole and this used linear programming (Jackson, 1971). The only other type of sector model that has been developed in the five-yearly assessment of the potential production of conifer roundwood in Britain, produced jointly by the Forestry Commission and private growers (HGTAC, 1984). The absence of any other forest sector models in Britain can probably be explained by the fact that they would be of relatively limited use as the price of wood product imports would have a major influence on any results and this would be an external variable over which the sector has no control and which would therefore be hard to predict.

Other countries such as Scandinavia and the USA have developed more sophisticated forest sector models. In preparing their most recent forest sector review (USDA, 1982), the US Forest Service used simulation models for projecting supply, demand and equilibrium prices for timber, plywood and pulpwood for both hardwoods and softwoods in nine different regions of the USA. Parameters in the model such as demand and supply elasticities were based on historical relationships.

Many forest sector models in the past have stopped at the 'forest gate' but the sector is now being more broadly defined to include marketing, manufacturing, transport and harvesting.

One such model is currently under development in New Zealand (Whyte and Baird, 1984), and Gane (1984) has developed a flexible model for use on a micro-computer in developing countries. These more broadly based sector models have the potential to make important contributions towards policies involving, for example, the development of forest industries. With the output of Britain's forests expected to double over the next 10 to 15 years, the development of regional models could help in, for example, comprehensive evaluations of the costs and benefits of alternative locations of forest industries as well as in the debate about the types of forest industries that are most appropriate within the United Kingdom.

In 1980 an ambitious project began involving the development of a Global Trade Model at the International Institute for Applied Systems Analysis (IIASA) in Vienna. This differs from the national forest sector models just referred to in that it is essentially concerned with the international trade in forest products between forest sectors in different countries. The preliminary model is based on six regions, (Northern Europe, the rest of West Europe, USA, Canada, Japan, and the rest of the World), and nine wood products. It is intended that the final version of the model will comprise 18 regions and 13 wood based commodities (Dykstra and Kallie, 1984) and will be completed by the end of 1984.

*Forest or Investment Models*

*Simulation Models.* These models can vary in size and complexity from those representing a single compartment or stand of trees to those for whole forests. They can be used for initial investment decisions, as well as for management decisions such as forecasting future volumes for harvesting and choosing between alternative silvicultural regimes.

One such model has been developed in New Zealand by the Radiata Pine Task Force at the Forest Research Institute in Rotorua (Holloway, 1982). The computer model, known as SILMOD, is able to simulate virtually any silvicultural regime (especially any combination of multiple pruning and thinning), on any rotation length (15 to 50 years), on almost any New Zealand site (flat to very steep). The model selects the most appropriate harvesting system and predicts transport costs to the sawmill and to the market after conversion. Sawmill types can be selected and conversion standards, methods, sawing patterns and some price assumptions can be altered at will.

In the United Kingdom a number of programs are currently being developed in the forest sector. Within the Economic Forestry Group we have developed two computer programs for a 64K micro-computer to help us with investment decisions of clients' properties.

The first program allows us to project timber volumes and income. The starting point for this program is a file containing data for each compartment of a property on species, date of planting, initial stocking, area, percent of the area productive, silvicultural management regime and hazard class. To forecast timber volumes and income the program uses Forestry Commission yield tables stored on a 0.5Mb floppy disc. Timber prices are entered separately to suit the particular circumstances. There are also facilities built into the program which allow any of the assumptions such as timber prices, yield classes or hazard classes to be changed so that their sensitivities can be assessed if required.

The second program is used for building up a detailed cash flow projection for periods up to 100 years and incorporates discounting, compounding and internal rate of return functions. Individual entries or columns of entries in the cash flow projection can be easily changed so there is again the opportunity to undertake any number of sensitivity analyses very simply. In this way it is possible to demonstrate the effect of investment decisions, to assess risk and to examine the importance of various assumptions made in preparing any projection.

It has become apparent from this sort of work that the existing yield tables published by the Forestry Commission as Booklet No. 48 have significant limitations. This is because of the restricted number of silvicultural regimes for which they have been prepared, the absence of yield tables for a number of spacings, and because forecasts will only be accurate if stands are managed exactly in accordance with the tables. The development of more flexible growth models for different species, for predicting volumes as well as assortment and grade outturn for stands under different silvicultural regimes and spacings, are considered to be important areas of further research.

*Optimising Models.* The principal type of model in this category, which has been used in North America and Scandinavia in particular for many years, is linear programming. Their major advantage is that they identify optimal solutions.

Linear programming has some potential applications in organisations with responsibility for the management of large forest areas which are all under one ownership such as the Forestry Commission (Grundy, pers. comm.). For example, it could be used to find the felling schedule which would maximise net present worth subject to an organisation's constraints on capital and manpower. However linear programming can have limitations in such a situation in that a considerable amount of aggregation is required if an optimal solution for an organisation is required. One consequence of this is that it is difficult to translate the results into detailed work programmes at the field level.

## VALIDATION OF MODELS

Having examined some of the different types of economic models that have been developed, it is also opportune to consider the need to validate before using them. This is an essential step once a new model has been completed but it is also important when using a model for the first time.

The process of validation involves assessing a model in relation to the objectives for which it was developed or the purposes for which it is to be used. Some models are designed for predictive purposes and others are designed to get a better understanding of a system. With most micro-economic and financial models, it is their predictive accuracy which is often most important, but in sector and national economic models the large number of variables and their aggregate nature means that greater emphasis may be placed on understanding the types of responses rather than the absolute level of those responses. Users should therefore be sure to ascertain the original purpose of an economic model before using it or accepting its results.

## CONCLUSIONS

Economic modelling is not a new concept but both the scope and extent of the activity have developed considerably with the introduction of computers.

Economic models do not provide a panacea for all types of decision making but they can often be a useful tool for examining the implication of different decisions prior to a decision being made. However it is important to recognise that economic models can be developed for a number of different purposes and that the results obtained from them will depend on those objectives and on the validity of the assumptions that have been made in the model.

There is considerable potential for the further development of economic models using computers at the stand and forest level as well as at the sector and global levels. It is important that these opportunities are taken up now so that well informed decisions can be made within the forestry sector to maximise its potential.

ACKNOWLEDGEMENTS

I wish to acknowledge with thanks the very helpful comments I have received on an earlier draft of the paper from the Editors and D.S. Grundy of the Forestry Commission.

REFERENCES

Club of Rome (1972) *Limits to Growth.* Earth Island Ltd, London.

Dykstra, D. and Kallio, M. (1984) A Preliminary Model of Production, Consumption and International Trade in Forest Products. *Working Paper WP-84-149.* International Institute for Applied Systems Analysis, Laxenburg, Austria.

Gane, M. (1984) Microcomputer Models for Forestry Sector Planning Paper to IUFRO Division 4. Thessaloniki, Greece.

Goulding, C.J. (1979) Validation of growth models used in forest management. *New Zealand Journal of Forestry,* **24**, 108-124

Holloway, J.S. (1982) Editorial Comment. *New Zealand Journal of Forestry,* **27**, 8-13.

HGTAC, (1984) Production Forecasts for Coniferous Roundwood. Supply & Demand Sub-Committee. *Forestry & British Timber*. Jan, 10-13.

Jackson, B.G. (1974) *Forest Products in the United Kingdom Economy.* Forestry Commission Bulletin, **51**.

Loasby, B.J. (1976) *Choice, Complexity and Ignorance.* Cambridge University Press.

Navon, D. (1973) Forest Management as seen by the Computer. Paper to Western Stand Management Committee. San Jose, California, 157-162.

Treasury (1981) *Economic Progress Report* **134**, 2.

US Dept. of Agriculture (1982) *An Analysis of Timber Situation in the United States 1952-2030.* Forest Resource Report, 23 USDA Washington.

US Government (1982) *The Global 2000 Report to the President.* Allen Lane, London.

Van Horn, R. (1969) Validation. *The Design of Computer Simulation Experiments.* Naylor, T.H., (ed.) Duke University Press, Durham, N.C.

Wardle, P.A. (ed,) *Operational Research & the Managerial Economics of Forestry.* Forestry Commission Bulletin, **44**. HMSO.

Whyte, A.G.D. and Baird, F.T. (1983) Modelling forest industry development. *New Zealand Journal of Forestry.* **28**, 275-283.

# FORESCO II—THE APPLICATION OF ECONOMIC MODELLING

Wm.C. HUMPHRIES, Jr. AND RON THOMPSON

*Forest Resource Consultants, Inc., Post Office Box H, Jeffersonville, Georgia 31044 USA*

## SUMMARY

Computers are rapidly becoming an important part of the forestry profession. This paper assumes no knowledge of forestry in the southeastern US. Therefore, a brief introduction to southern US forestry is provided including a description of ownership patterns, forest conditions, markets for standing timber and typical forest management practices normally used. The paper concludes by illustrating how the computer and software developed by Forest Resource Consultants, Inc. aids our forest managers in developing forest management and harvesting decisions, with emphasis given to economic as well as silvicultural considerations.

## INTRODUCTION

The aims of our paper are threefold. First, we will provide some general background to forestry in the southern pine region of the US. Second, we will outline ownership patterns, forest cover types, stocking conditions, timber markets, and general schemes of management. Finally, we will show the importance of the computer in solving some typical economic problems posed in forest management.

## OWNERSHIP

According to government statistics, the southeastern states (Figure 1) contain approximately 188,000,000 acres (7611 k ha) of commercial timberland. I would further estimate that approximately 80 percent of this area could be classed as southern pine.

Approximately 70 percent of the land belongs to the private owners, mostly individuals or families (Lewis, 1978). Twenty percent of the land is owned by private forest industry and the remaining 10 percent is owned by various state, federal, or municipal agencies. Much of the federal government's land in the southeast is located in the mountainous regions. Therefore, most of the land stocked with southern pine, the principal group of species having commercial value, is privately owned.

## FOREST CONDITIONS

There are four major southern pine species: loblolly (*Pinus taeda*), slash (*Pinus elliottii*), longleaf (*Pinus palustris*) and shortleaf (*Pine echinata*). The lumber produced by each species is of similar quality and has the same commercial value. About 80 percent of our timber is in natural stands that have developed on abandoned farmland through natural regeneration. The remaining 20 percent is pine plantations, most of which have been established within the past 25 years. While planted stands are mostly pure pine, our natural stands vary from pure pine to mixtures of pine and hardwood. Most hardwoods growing on pine sites are of little commercial value and are dealt with as undesirable competition. Markets and prices for hardwood in the south are relatively weak as compared to pine. Hardwoods of commercial value include various species of oak (*Quercus*), yellow poplar (*Liriodendron tulipifera*), sweetgum (*Liquidambar stryraciflua*), elm (*Ulmus*) and maple (*Acer*) and only command about 30 to 50 percent of pine timber prices. Therefore, efforts are concentrated primarily on the establishment, management and marketing of pine.

## MARKETS FOR STANDING TIMBER

In most areas of the south, a three-product market exists including pulpwood, chip-n-saw (small sawtimber) and large sawtimber. In most cases, a single company bids for the tract or sale area as a whole and standing on the stump, then cuts, separates and hauls the various products to separate markets. Sometimes a single buyer processes all three products.

There is a high degree of stability among markets for standing timber. It is seldom that, on a constant dollar basis, price fluctuations exceed 10 to 15 percent for a specific market area within a 12 to 24 month period. However, prices for each product do vary between areas of the south or within a state. In the state of Georgia, three distinct market areas exist. Consider the following:

| | North Georgia | Middle Georgia | South Georgia |
|---|---|---|---|
| Pulpwood | $15/cord[1] | $20/cord | $30/cord |
| Chip-n-saw | $35/cord | $40/cord | $45/cord |
| Sawtimber | $50/cord | $60/cord | $65/cord |

[1]One cord=approximately 90 cubic feet of wood and bark. One cubic metre=35.31 cubic feet. Exchange rate: 1 dollar=1.22 pounds sterling. Using these rates, 20 dollars per cord=6.45 pounds per cubic metre.

As we will later see, recognising product price variation is important in determining the rotation length and frequency of thinning. Generally speaking, the higher the price for small products, such as pulpwood, the earlier a stand reaches financial maturity and the importance of thinnings is reduced.

## SILVICULTURAL PRACTICES

Although management procedures and cultural practices are gradually becoming more intensive, management in the southeast US is extensive compared to British forestry.

A normal tract of 1,000 acres, primarily natural growth, may consist of 20 to 40 separate stands with varying ages, levels of stocking, degree of hardwood competition and site quality conditions. These variations create multiple alternatives for management and harvesting and varying years of financial maturity.

A typical scheme of management for natural stands might be as follows:

Age 2-5: Assess establishment results and determine the need to control undesirable scrub hardwood competition.

Age 15-20: A control burn using the backfire method may be prescribed to control undesirable hardwood competition.

Age 20-25: Light improvement thinning removing only pulpwood, possibly followed in two or three years by another burn.

Age 30-35: Second improvement thinning removing pulpwood plus some small sawtimber to improve spacing and improve crown growth.

Age 40-45: Remove all trees except a few dominant crop trees which will serve as a seed source for re-establishing the new crop.

Age 47-52: Remove seed trees—hope adequate natural regeneration has occurred. If not, clear land with a tractor and replant.

Obviously, this is a theoretical model. There often are variations to this scheme that may include fewer thinnings, with final harvest at age 35, except for the remaining seed trees.

A typical pine plantation management scheme might include the following:

Year 0: Mechanically clear the site and mechanically plant approximately 750 trees per acres.

Year 12-18: Prescribe burn to control the hardwood understory.

Year 18-22: First thinning, possibly followed by controlled burning.

Year 22-26: Second thinning, possibly followed by controlled burning.

Year 28-35: Final harvest, clearfell all volume, and begin site preparation (clearing) and establishment of the next crop.

Intensive practices such as fertilisation are employed in certain situations. Applying phosphate on wet, waterlogged sites is a common practice in the flat coastal areas. There is also some experimental work with nitrogen fertilisation but this practice is not widely accepted at this time.

By contrast with British practice, fire is an important management tool. It is our least expensive and the most environmentally acceptable form of hardwood and brush control provided good smoke management practices are employed. Southern pines have a relatively thick bark with good insulating qualities. Therefore, no pine damage usually occurs if burning is under prescribed weather conditions.

## TRENDS IN SOUTHERN FOREST MANAGEMENT

Due to a plentiful supply and low prices, efforts prior to 1950 were concentrated on exploitation or harvesting with a limited amount of effort devoted to establishment and management techniques. With increasing timber prices and predicted shortages of timber the pulp and paper industry in the southeast has been diligent in its efforts to improve the management and establishment of pine. On the other hand, the private non-industrial sector, except for a small percentage of owners, has continued to be a relatively poor steward.

Industry and private foresters have concentrated primarily upon the silvicultural aspects of forestry or 'gardening' and have devoted less effort towards forest economics. This is probably typical of most young professions. In the past, forestry schools have concentrated on training foresters in silviculture and devoted little emphasis to the economic and financial considerations of timber production. Even basic research is often conducted and published in a form that does not readily lend itself to an economic evaluation. For example, there have been intensive studies on pine growth response after controlling unwanted hardwood species but only a minimum of effort is devoted to the economics of such practices (Loyd *et al* 1978).

In the past, some forest industries and most non-industrial landowners viewed their timberland as a capital reserve to be raided when wood was needed to feed a mill or to raise cash to supply personal needs. Today, particularly in the past five to ten years, many non-industrial landowners have begun to take a different view of their

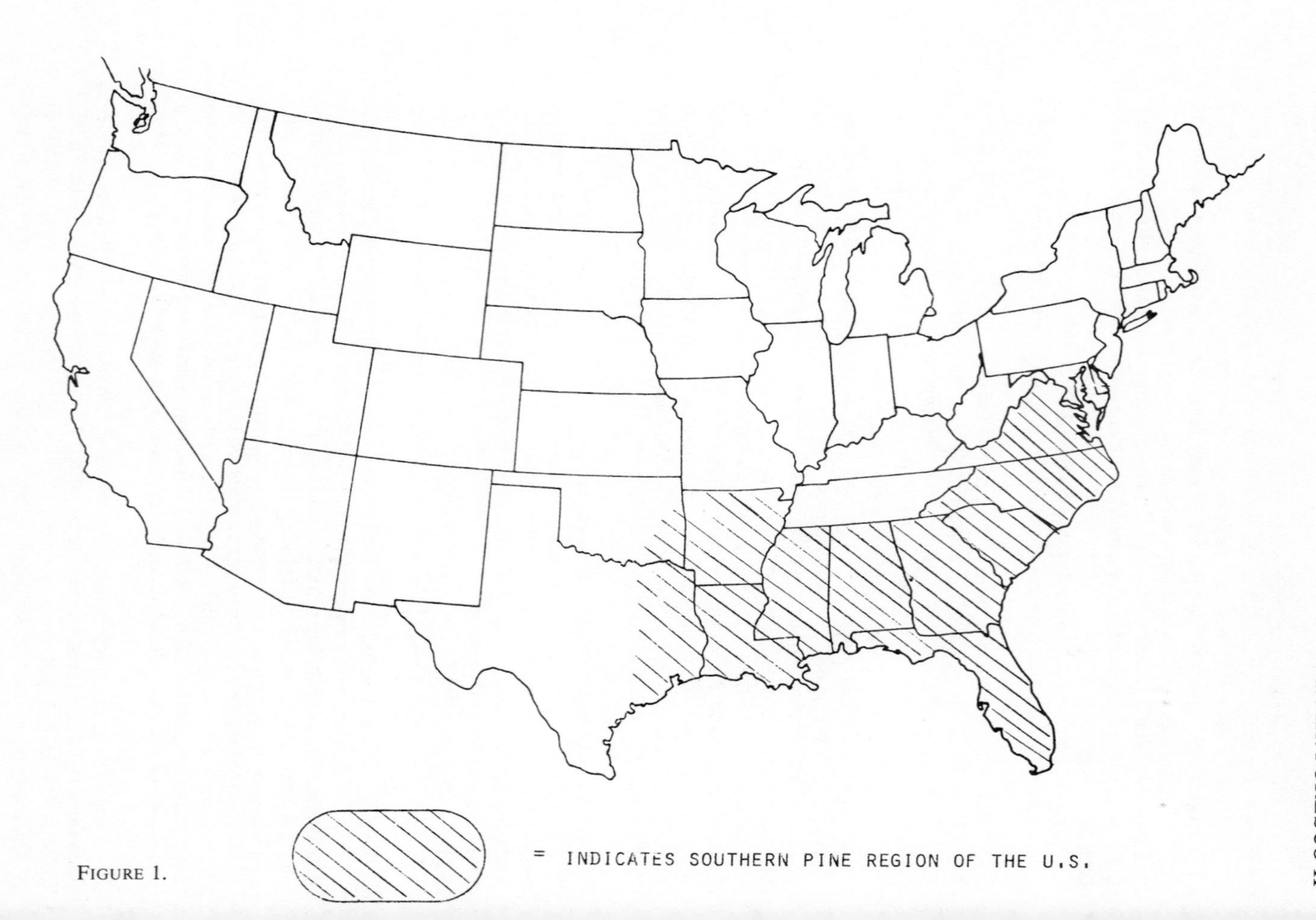
= INDICATES SOUTHERN PINE REGION OF THE U.S.

FIGURE 1.

timberland. Some industries now consider their timberland as a separate profit centre—a separate investment.

This emphasis on timberland as an investment in its own right has caused many foresters to become more interested in the economic aspects of timber management and marketing. At least one college, the University of Georgia School of Forest Resources, now offers advanced degrees in forest business management. Also, numerous continuing education seminars are being offered by various colleges in the South in the area of forest finance, economics and investment analysis.

## FORESCO II

During the past three years, our firm has invested heavily in personnel training and software development in an effort to improve our ability to address the economic issues of timber management.

What are some of the typical economic questions that apply to forest management and the marketing of timber? Consider the following:

How much can be spent on site preparation and planting considering different site qualities, future yields and management costs?

How much can one pay for a timber tract given an inventory of timber, land conditions and market conditions?

Are physical maturity and economic maturity synonymous? Foresters, relying upon silvicultural training, have considered that a crop should be harvested when the current annual growth falls below the mean annual growth. This precept gives no consideration to value, merely to volume growth. We now know that economic maturity occurs earlier than physical maturity under most market conditions.

Is a management regime for the south Georgia market area the same for the middle Georgia market? Price structures are different; therefore, rotation age and thinning requirements do vary between the two markets.

How should a typical stand be harvested? Should it be thinned once, twice, three times or more and how intensively should the stand be thinned at each interval? At what age should thinnings occur? Should the final harvest be made at age 20, 30 or 40 years of age?

Such questions cannot be answered or examined without the aid of a computer. Given the complexity of the relationships involved, our firm has developed a software package resulting in a forest management system named FORESCO II.

FORESCO II is a data base and forest data processing system that simulates volume growth, projects stumpage prices, and analyses the economics of various thinning and harvesting alternatives using either internal rate of return or present value as the measure of profitability.

A flow chart (Figure 2) illustrates the process from field level to the final reports. The process begins with an owner-manager conference where general owner objectives and requirements are established. The system requires the forest manager to provide input and review at various stages of development, thereby providing assurance that the final management plan shows that personal judgement and control prevailed. The final report is a forest management plan that sets forth annual work plans for a ten year period with further cash flow figures for a full thirty year period. The system further provides annual reports including: a summary of the previous year's activity; income and expense statements; an annual depletion report for the accountant; an annual fair market value report (price property would bring if

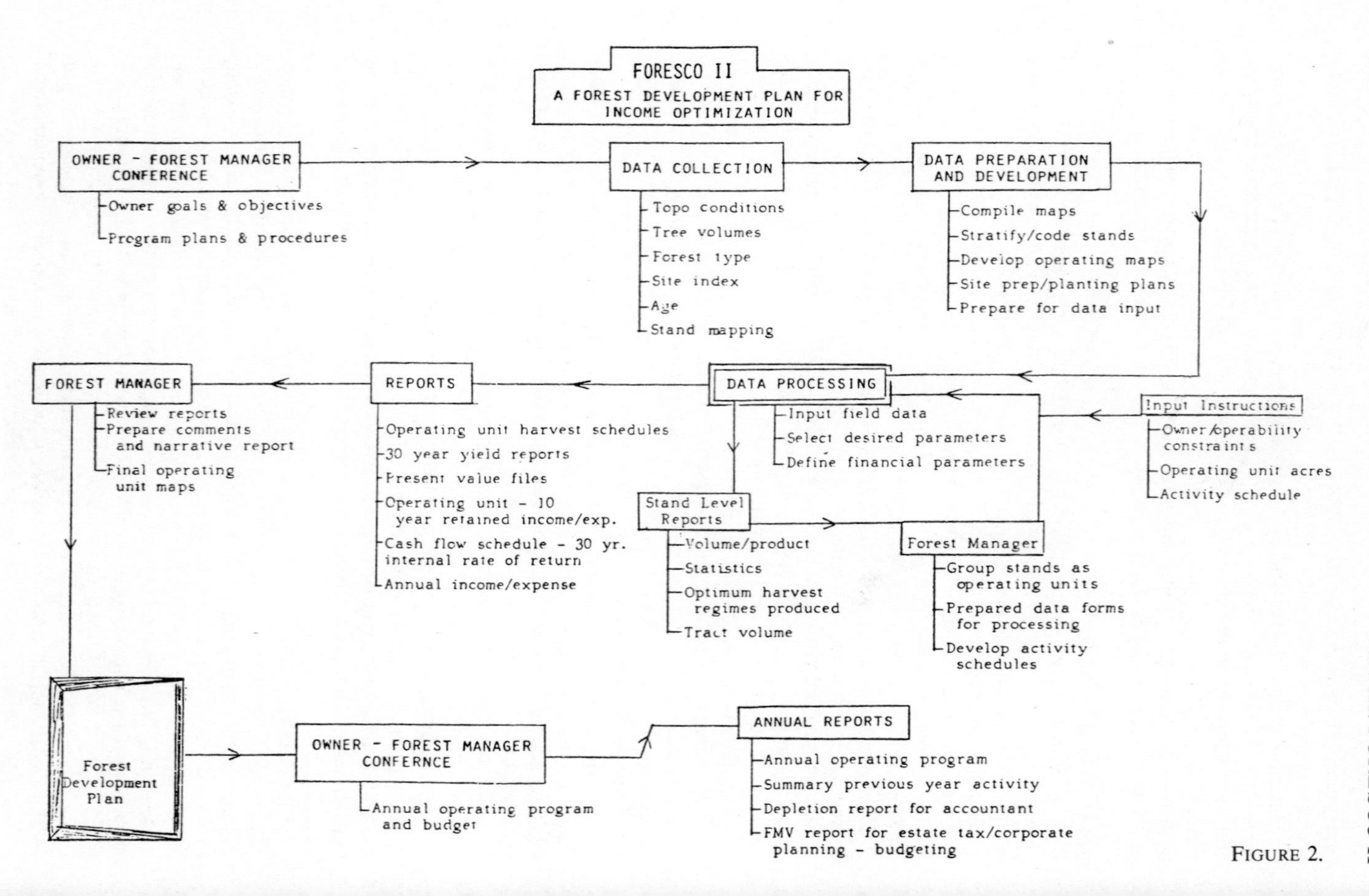
FORESCO II
A FOREST DEVELOPMENT PLAN FOR INCOME OPTIMIZATION
OWNER - FOREST MANAGER CONFERENCE
Owner goals & objectives
Program plans & procedures
DATA COLLECTION
Topo conditions
Tree volumes
Forest type
Site index
Age
Stand mapping
DATA PREPARATION AND DEVELOPMENT
Compile maps
Stratify/code stands
Develop operating maps
Site prep/planting plans
Prepare for data input
FOREST MANAGER
Review reports
Prepare comments and narrative report
Final operating unit maps
REPORTS
Operating unit harvest schedules
30 year yield reports
Present value files
Operating unit - 10 year retained income/exp.
Cash flow schedule - 30 yr. internal rate of return
Annual income/expense
DATA PROCESSING
Input field data
Select desired parameters
Define financial parameters
Input Instructions
Owner/operability constraints
Operating unit acres
Activity schedule
Stand Level Reports
Volume/product
Statistics
Optimum harvest regimes produced
Tract volume
Forest Manager
Group stands as operating units
Prepared data forms for processing
Develop activity schedules
Forest Development Plan
OWNER - FOREST MANAGER CONFERNCE
Annual operating program and budget
ANNUAL REPORTS
Annual operating program
Summary previous year activity
Depletion report for accountant
FMV report for estate tax/corporate planning - budgeting

FIGURE 2.

offered for sale) and updated cash flow schedules for estate or corporate planning and budgeting.

To illustrate how the computer and this program has aided us in evaluating the economics of various harvesting alternatives, we will consider the following example of a stand located within one of our tracts in the middle Georgia area. The object of this exercise is to find the harvest year in which internal rate of return is maximised.

Stand parameters developed from field notes and the processed inventory are as follows:

| | | |
|---|---|---|
| Acres: 120 | Pine pulpwood: | 3.9 cds/acre (24 $m^3$/hectare) |
| Species: loblolly | Pine chip-n-saw: | 4.0 cds/acre (25 $m^3$/hectare) |
| Site index: 1/90 | Pine sawtimber: | 11 cds/acre (67 $m^3$/hectare) |
| Age: 35 | Density: | High |

Timber prices, cost and other financial parameters used for developing the best harvest regime are shown at the top of Figure 2.

The evaluation begins by matching the stand to the appropriate growth and yield model. For clarification, a stand is an area of timber and land that is relatively homogeneous with regards to species, cover type, stocking or density, age and soil productivity (site index). Yield models are of two general categories—whole stand (projection of cubic foot volume) and diameter distribution models (projection by diameter classes).

Once the model is selected, future volumes under various harvesting alternatives are projected using initial characteristics of the stand collected from a field inventory. The stand's internal rate of return is determined for each year in the planning period of 30 years based upon clearfelling and including the value of any previous thinnings that proved feasible.

The year of financial maturity for the existing stand is also influenced by the value of the subsequent crop. When the existing stand fails to perform as well financially as the new crop is capable of performing, the existing stand is deemed to be financially mature. However, if the owner is only concerned with the existing crop, as may be the case with certain special lease arrangements, the year of maturity may be different. Hence, owner objectives are an important consideration in the development of a forest management plan.

Various thinning alternatives are evaluated but under certain constraints. For example, a five year thinning response must be allowed before subsequent thinnings or final harvest can occur and a minimum of five cords per acre for thinning must be met. Once these two constraints are satisfied, three intensities of thinnings are evaluated. These intensities are:

1. Light thinning—remove volume until residual basal area stocking is equivalent to the site index plus 5. Example: If site index is 80 then add 5 points, so remove volume until a residual basal area stocking is 85.
2. Medium thinning—(Residual stand density = site index − 5.)
2. Heavy thinning—(Residual stand density = site index − 10).

When thinning regimes are included the number of harvesting alternatives increases dramatically in the later years of the planning period. Although there could be literally thousands of alternatives to evaluate, we have narrowed our alternatives to about 10 or 15 harvest regimes. Without the aid of the computer the evaluation of clearfellings alone, without intermediate thinnings for each year in the planning

FIGURE 3.

Financial maturity

| | | | | | |
|---|---|---|---|---|---|
| Client: | Wilson | State: | Georgia | Stand No.: | 12 |
| Owner: | Wilson | County: | Twiggs | Acres: | 120 |
| Tract: | Homeplace | MGMT Unit: | 1 | Strata: | P-3-35-H |

Financial parameters

Stumpage values ($/cord)

| Species | Product | Existing stand thinning | Existing stand harvest | Regenerated stand thinning first | Regenerated stand thinning second | Regenerated stand harvest | Stumpage Appreciation rates | |
|---|---|---|---|---|---|---|---|---|
| Pine | P/W | 18.00 | 20.00 | 18.00 | 18.00 | 20.00 | 0.00% | |
| | CNS | 34.00 | 40.00 | 18.00 | 34.00 | 40.00 | 0.00% | 0.00% |
| | SWT | 34.00 | 60.00 | 18.00 | 34.00 | 60.00 | 0.00% | Infl rate |
| Hardwood | P/W | | 3.00 | | | | 0.00% | 0.00% |
| | SWT | | 12.00 | | | | 0.00% | |

| | | | |
|---|---|---|---|
| Sales commission | Site prep & planting | Ad val tax: $3.00 | Lease inc: $0.00 |
| Thinnings 10.00% | Method: Chop/burn | Mgmt fee: $2.00 | Mic inc: $0.00 |
| Harvest 5.00% | Cost ($/AC) $165.00 | Misc exp: $1.00 | |

Optimum harvest regime for each harvest year

| | Harvest year | Age | Regime number | IRR | |
|---|---|---|---|---|---|
| | 1984 | 35 | 1000000 | 3.35% | |
| | 1985 | 36 | 2000000 | 3.44% | |
| | 1986 | 22 | 3000000 | 3.51% | |
| | 1987 | 23 | 4000000 | 3.55% | |
| | 1988 | 24 | 5000000 | 3.57% | 1% Planning Window |
| Financial Maturity → | 1989 | 25 | 6000000 | 3.59% | |
| | 1990 | 26 | 7000000 | 3.58% | |
| | 1991 | 27 | 8000000 | 3.57% | |
| | 1992 | 28 | 9000000 | 3.54% | |
| | 1993 | 29 | 10000000 | 3.51% | |
| | 1994 | 30 | 11000000 | 3.46% | |
| | 1995 | 31 | 12000000 | 3.41% | |
| | 1996 | 32 | 13000000 | 3.36% | |
| | 1997 | 33 | 14000000 | 3.31% | |
| | 1998 | 34 | 15000000 | 3.25% | |
| | 1999 | 35 | 16000000 | 3.19% | |
| | 2000 | 36 | 17000000 | 3.13% | |
| | 2001 | 37 | 18000000 | 3.07% | |
| | 2002 | 38 | 19000000 | 3.01% | |
| | 2003 | 39 | 20000000 | 2.95% | |
| | 2004 | 50 | 21000000 | 2.89% | |
| | 2005 | 41 | 22000000 | 2.83% | |
| | 2006 | 42 | 23000000 | 2.77% | |
| | 2007 | 43 | 24000000 | 2.71% | |
| | 2008 | 44 | 25000000 | 2.66% | |
| | 2009 | 45 | 26000000 | 2.60% | |
| | 2010 | 46 | 27000000 | 2.55% | |
| | 2011 | 47 | 28000000 | 2.50% | |
| | 2012 | 48 | 29000000 | 2.45% | |
| | 2013 | 49 | 30000000 | 2.40% | |

Regime number: 10000000

| 10 | 00 | 0 | 00 | 0 |
|---|---|---|---|---|
| HARVEST YR | 2ND THIN YR | INTENSITY | 1ST THIN YR | INTENSITY |

Prepared by Forest Resource Consultants, Inc.

period, would take an estimated 10 hours to complete using a hand-held calculator. Using a computer we are able to complete the same evaluations and receive a printed report in approximately six minutes.

Once all evaluations have been completed, a stand report (Figure 3) indicates the rate of return for each year of harvest and it further highlights a 'planning window' and by an asterisk denotes the exact year of financial maturity. In our example, financial maturity is achieved by clearfelling in 1989 without intermediate thinnings. In this example, no thinnings proved practical as indicated by the regime code number. From a practical point of view the rate of return is affected very little if harvesting occurs between 1987 and 1992 which provides some latitude to take advantage of market conditions and cash flow adjustments. An identical report is generated for each stand in the tract. The report for each stand is made available to the forest manager for use in organising management or operating units, and eventually the computer compiles a final operating unit harvest schedule and cash flow schedule (Figure 4) for the entire ownership.

This cash flow report (Figure 4) allows the forest manager to evaluate a prospective purchase of timberland in terms of future cash flow on a constant dollar basis and with various assumed inflation rates. Other reports produced by FORESCO II will allow the forester to analyse a timber investment on an after-tax basis with and without leveraging (financing).

## CONCLUSIONS

It has been my observation that one of the greatest obstacles to improved forest management and more timberland investment has been the inability of foresters to make a reasonably reliable presentation of cash flow expectations and the inability to express financial yields on a before-and-after tax basis.

The computer, through FORESCO IIsoftware, allows us to provide a landowner or prospective owner with the ability to analyse a timber investment and compare it with alternatives. More intensive management and improved economic evaluations of management alternatives are imminent. Rather than purchase more land—a capital intensive approach—industrial landowners and many private owners are seeking ways to manage their existing lands more productively. The micro-computer will obviously play an important role by providing the ability to simulate the growth and development of the forest under various silvicultural systems and economic assumptions. The computer industry is doing its part to provide us with reliable hardware at a reasonable cost. The challenge before us as foresters is to develop reliable software and to learn to use it effectively.

## REFERENCES

Bailey, R.L.; Pienaar, L.V.; Shiver, B.D.; Rheney, J.W. (1982) Stand Structure and Yield of Site-Prepared Slash Pine Plantations. *University of Georgia, College of Agriculture Experiment Station, Research Bulletin No. 291.*

Burk, T.E.; Burkhart, H.E. and Cao, Q.V. (1984) Powthin, Version 1.0. School of Forestry and Wildlife Resources, Virginia Polytechnic Institute and State University.

Clawson, M. (1979) The Economics of US Non-Industrial Private Forests. *Resources for the Future Research Paper No. R-14.*

Foster, B.B. (1982) Four Horsemen of the NIPF Apocalypse. *J. of Forestry.*

Gunter, J.E. and Haney H.L. (1984) Essentials for Forestry Investment Analysis.

Humphries, W.C., Jr. (1984) The Economies of a Pine Plantation Investment. (Unpublished).

PROPERTY CASH FLOW SUMMARY
(ALL VALUES ON CONSTANT DOLLAR BASIS)

CLIENT Wilson COUNTY, GEORGIA
TRACT Homeplace TOTAL ACRES 3733

| YEAR | ACTIVITY | MGMT. UNIT | INCOME PERIODIC | INCOME ANNUAL | EXPENSES PERIODIC | EXPENSES ANNUAL | NET INCOME |
|---|---|---|---|---|---|---|---|
| 1984 | LIQUIDATION VALUE PLUS 1984 INCOME LESS 1984 EXPENSE........... | | | | | | -3297734 |
| 1985 | | | 685800 | 0 | 44914 | 21465 | 619421 |
| 1986 | S | | 687900 | 0 | 143108 | 21465 | 523327 |
| 1987 | E | | 139400 | 0 | 160057 | 21465 | -42122 |
| 1988 | E | | 79500 | 0 | 105111 | 21465 | -47076 |
| 1989 | | | 234800 | 0 | 64989 | 21465 | 148346 |
| 1990 | A | | 396600 | 0 | 46725 | 21465 | 328410 |
| 1991 | C | | 334800 | 0 | 50140 | 21465 | 263195 |
| 1992 | T | | 0 | 0 | 38072 | 21465 | -59537 |
| 1993 | I | | 0 | 0 | 32200 | 21465 | -53665 |
| 1994 | V | | 0 | 0 | 672 | 21465 | -22137 |
| 1995 | I | | 0 | 0 | 0 | 21465 | -21465 |
| 1996 | T | | 192600 | 0 | 10880 | 21465 | 160255 |
| 1997 | Y | | 0 | 0 | 0 | 21465 | -21465 |
| 1998 | | | 0 | 0 | 8300 | 21465 | -29765 |
| 1999 | S | | 0 | 0 | 0 | 21465 | -21465 |
| 2000 | C | | 0 | 0 | 0 | 21465 | -21465 |
| 2001 | H | | 0 | 0 | 0 | 21465 | -21465 |
| 2002 | E | | 0 | 0 | 0 | 21465 | -21465 |
| 2003 | D | | 0 | 0 | 0 | 21465 | -21465 |
| 2004 | U | | 0 | 0 | 0 | 21465 | -21465 |
| 2005 | L | | 62160 | 0 | 6216 | 21465 | 34479 |
| 2006 | E | | 96880 | 0 | 9688 | 21465 | 65727 |
| 2007 | S | | 65072 | 0 | 6507 | 21465 | 37100 |
| 2008 | | | 32256 | 0 | 3226 | 21465 | 7566 |
| 2009 | | | 15344 | 0 | 1534 | 21465 | -7655 |
| 2010 | | | 20608 | 0 | 2061 | 21465 | -2918 |
| 2011 | | | 29680 | 0 | 2968 | 21465 | 5247 |
| 2012 | | | 27664 | 0 | 2766 | 21465 | 3433 |
| 2013 | 1/ | | 0 | 0 | 0 | 21465 | -21465 |
| 2014 | FINAL VALUES | 4619035 | 0 | 0 | 92381 | 21465 | 4505190 |

1/ SEE ATTACHED PAGE "FINAL VALUES OF REGENERATED STANDS".

THE INTERNAL RATE OF RETURN FOR THE ABOVE CASH FLOW PRESENTATION IS 3.24%

| | TIMBER | LAND | TOTAL |
|---|---|---|---|
| CURRENT LIQUIDATION VALUE (1984)...................... | 2400134 | 1116925 | 3517059 |
| PROJECTED VALUE AT END OF PERIOD (2013).............. | 3502110 | 1116925 | 4619035 |
| NET CASH FLOW, FOR PERIOD (1984-2013)................................. | | | 1624606 |

PREPARED BY FOREST RESOURCE CONSULTANTS, INC.

FIGURE 4.

Lewis, J.A. (1978) Landownership in the United States. *US Dept. of Agriculture, Agriculture Information Bulletin 435.*

Loyd, R.A.; Thayer, A.G. and Lowry, G.L. (1978) Pine Growth and Regeneration Following Three Hardwood Control Treatments. *Southern J. of Applied Forestry.*

Saucier, J.R.; Phillips, D.R. and Williams, J.G., Jr. (1981) Green Weight, Volume, Board-Foot and Cord Tables for the Major Southern Pine Species. *Georgia Forestry Commission Research Paper, No. 19.*

US Dept. of Agriculture (1982) A Guide to Federal Income Tax for Timber Owners. *Forest Service, Agriculture Handbook No. 596.*

Virginia Polytechnic Institute and State University (1984) Diameter Distributions and Yields of Natural Stands of Loblolly Pine. *School of Forestry and Wildlife Resources, Publication No. FWS-1-84.*

Wall, B.R. (1952-77) Trends in Commercial Timberland Area in the United States by State and Ownership, With Projections to 2030. *US Dept. of Agriculture, Forest Service, General Technical Report WO-31.*

# DESCRIPTION OF AN INTEGRATED FOREST MANAGEMENT SOFTWARE PACKAGE DEVELOPED FOR BRITISH COLUMBIA

ALAN MOSS[1]

*Alan Moss and Associates Ltd. P.O. Box 400, Winfield, British Columbia, Canada VOH 2CO*

SUMMARY

The paper describes a computer package for forest management planning use. This package includes programs for data capture from maps, data sorting, and for simulation of yields resulting from different management practices. It has been utilised in three Tree Farm Licence applications to the British Columbia Ministry of Forests with data inputs and analyses of approximately 160,000 hectares. The package may be expanded in several directions or modified to meet specific needs and preferences of management.

## INTRODUCTION

The purpose of this paper is to describe in brief a package of computer software developed to meet forest management planning needs in British Columbia. It has been specifically employed in data capture, data processing and management analyses of 160,000 hectares of Interior Dry Belt[2] forest for purposes of Tree Farm Licence[3] applications to the British Columbia Minstry of Forests. Results were technically acceptable.

## GENERAL DESIGN CONSIDERATIONS

The series of programs may be used individually or in combination. The package may be expanded or modified to meet specific management needs and preferences. Management options for data forms and data processing are incorporated. The management simulation program can be used interactively but is designed for use by professional foresters, and requires judgement in assessing its outputs.

Ease of data access and amendment to reflect the changing forest and incorporate improved research knowledge have been a principle of development. An emphasis was placed on speed of forest data entry and economic use of time of personnel of various qualifications required to operate the system. At the present time the package runs on a desk-top mini-computer of 96K bytes suitable for field management offices.

[1]The Institute records with regret the death of Dr Moss in the period between the conference and the publication of the proceedings. The final draft of his paper was in preparation at the time of his death and has been completed by the Editors. We apologise to the reader for any confusion caused by our lack of familiarity with British Columbian conditions.

[2]Editors' footnote: This has been interpreted as being synonymous with the Interior Douglas-fir zone of Krajina *et al.* (1982).

[3]For details of Tree Farm Licences see Taylor, G.W. (1975), *Timber: History of the Forest Industry in B.C.*, J.J. Douglas Ltd, Vancouver, pp. 162-163.

TABLE 1. Area summary for species group FIR

| Age Class | Site | | | | Total |
|---|---|---|---|---|---|
| | Good | Medium | Poor | Low | |
| NSR | 0 | 68 | 0 | 0 | 68 |
| 1 | 2 | 160 | 0 | 0 | 162 |
| 2 | 0 | 10 | 0 | 0 | 10 |
| 3 | 0 | 4 | 0 | 0 | 4 |
| 4 | 0 | 195 | 0 | 0 | 195 |
| 5 | 0 | 32 | 0 | 0 | 32 |
| 6 | 16 | 4 | 0 | 0 | 20 |
| 7 | 0 | 0 | 0 | 0 | 0 |
| 8 | 66 | 118 | 0 | 0 | 184 |
| 9 | 0 | 0 | 0 | 0 | 0 |

Note: NSR is not satisfactorily stocked and requiring regeneration.

## DESCRIPTION OF PROGRAM

### *Map capture program*

This program permits the user to digitise the boundaries of areas ('type islands') corresponding to different forest cover types from a forest cover map, according to a 'menu' organised by the user. Areas of type islands are computed with a record of area deductions for lakes, roads and a visual centroid (map reference of the type island) recorded. In addition to areas and location, a number identifying the stand, forest cover map label information, lengths of rights-of-way by uses and classes are recorded. Desired comments such as wildlife use and environmental sensitivity are also included. Checks are incorporated to eliminate or control digitising error.

### *Data sorting programs*

Table 1 gives a breakdown of the captured data for the Fir species group, listing stands by age class and site quality. Similar tables can be provided for other species groups (e.g. pine and spruce) together with details of areas subject to management constraints (e.g. Environmental Protection Areas) and listings of unproductive ground. All this information can then be summarised as shown in Table 2. (p.124)

Captured data are sorted as desired using these programs. A 'universal' sort program is employed when a large data set is to be analysed, while other programs are used to save time when searching for a few items. The latter may be employed for example in answering day-to-day management questions such as 'What area of pine aged 15-20 years is classified as environmental protection forest and where are the stands located?'. The 'universal' sort of program is employed to organise data for the management simulation program 'SYLVA' described below.

### *Avline data program*

B.C. Ministry of Forests Cover Maps each have a set of separate tables giving gross and net average volumes for stands of a given species composition, age, height, stocking and site quality. The Avline program applies this information in combination with results from the data sorting program to provide estimates of standing volume.

### *Volume compilation program*

This program utilises results from a ground inventory to compute tree, stand and overall area volumes. It can be used to modify or replace Avline data to increase the accuracy of volume estimates.

*'SYLVA' management simulation program*

The program has been used to simulate the management practices of the forest products industry in B.C. although it can also be applied to public management. Data have been sorted into commercial species groups or groups in which a single species is dominant, so that the possibilities of supply of a given balance of species and the time over which this supply may be maintained can be identified. Each species group is analysed separately by site quality class since silvicultural prescriptions normally vary between classes. Other data breakdowns may be employed to allow for special prescriptions for environmentl protection forest and special wiildlife areas.

Table 3 gives an example of a program run using the data listed in Table 1. Areas are reduced to allow for Environmental Protection Area deletions (there are minor rounding errors in the figures). Items 1-10 are mandatory inputs by the user using figures based on professional knowledge, apart from item 5 which is calculated from Table 1. Where roads are concerned (item 6), the percentage input provided for roadbuilding is deducted from the productive area before yields are predicted. Exisiting roads will already have been allowed for during the data sorting stage (see Table 2). The availability of improved seed (items 11-12) is an optional input. Other optional inputs include the frequency of fertilisation and the percentage improvement expected in growth.

Thinning parameters (item 13) are based on growth rates given for thinned stands in Forestry Commission Management Tables (Hamilton and Christie, 1971), but reduced by 5 percent. These tables need to be used because there are currently insufficient data available for British Columbia. The number of thinnings, the ages at which they occur, and the percentage of gross volume to be removed (stand depletion) are specified by the user. First thinnings are considered to be noncommercial but are restricted to overstocked stands. An optional input constraint is to prevent simulation of thinning in stands over a certain age where windthrow might be expected (eg spruce). A graphics display presents the operator with the growth curve for the particular forest and the effects of the proposed thinnings upon it. This allows the user to correct the proposed regime.

Any updating of data is carried out by the computer as a result of the operator entering the number of years since the last revision. It is presumed that major changes due to logging or forest fires will have been updated on the maps prior to map capture.

After items 1-14 have been entered, initial outputs are then produced by the program as shown in Table 3, items 15-20. These preliminary outputs are designed as a guide to the user and do not affect the actual simulation which is governed by entering items 21-24. Sustained annual felling yield (item 19) is based upon a sustained yield formula. Thinning yields are estimated by following the prescription entered previously (item 13) and averaging yields.

After reflection, the user then enters the chosen value for items 21-24. The most important figure is that for item 22 which regulates the simulation process. The value used may reflect the raw material requirement of a timber-using industry or it could indicate the potential supply from the forest over a given time period. The figure may or may not allow for a sustained yield strategy. Similarly, the figure for the maximum annual thinning area (item 21) will be whatever the user feels to be achievable given his resources. The chosen value may seek to avoid excessive year-to-year fluctuations in thinning yields.

Finally, items 24 and 23 respectively set the duration of the simulation and the intervals at which are to be produced.

## SIMULATION PROCESS

The computer starts at year 1 and scans the data calculating the effect of operations. Data are changed accordingly and one year's growth is added on before the program carries on to the following years. The simulation thus proceeds stepwise until the final year (item 24) has been reached. Decisions on choosing felling and thinning areas will have been governed by stand age. Area-volume checks are output as the simulation proceeds (Table 4) at the predetermined interval (item 23). Finally a graphical presentation of annual yields (Figure 1) and a summary of operations for the first 20 years (Table 5) are produced. Other summaries could be output on an aggregate basis such as predicted yields for one species group on all sites, or for all species and sites combined.

The simulation can also be used to predict the allowable cut in each of the first five years of operations and the anticipated sustained annual yield.

## OTHER PROGRAMS

One program accepts sorted data and produces a graphical presentation of age class distribution as compared to a normal forest. It functions on individual species groups by site class or any aggregate of data up to all species on all sites. A further program produces maps from map capture data which convey a clear visual impression of locations of mature stands by age class and dominant species. A letter indicates the dominant species, with its colour showing the age class and its size indicating the area of the stand.

## DISCUSSION

The package provides a system by which managers or planners may evaluate a given forest area and plan its management regime. Costs are produced based on current values. For British Columbia purposes stumpage rebates for which some operations qualify are calculated and checked against a Ministry of Forests formula which limits the rebates relative to annual yield(s). There is much opportunity for further expansion of the software package and its adaptation to other situations. Integration with conversion management software is one possibility and a more sophisticated economic analysis is probably desirable. To date, the system has been developed in response to actual demands and has found good acceptance in the situations to which it has been applied.

## REFERENCES

Hamilton, G.J. and Christie, J.M. (1971) *Forest Management Tables (Metric).* Forestry Commission Booklet No. 34. HMSO. London. 201pp.

Krajina, V.J.: Klinka, K. and Worrall, J. (1982) *Distribution and Ecological Characteristics of Trees and Shrubs of British Columbia.* University of British Columbia, Faculty of Forestry. 131pp.

TABLE 2. Summary of areas in hectares

| **Productive Forest Land** | | | | | | **Total** |
|---|---|---|---|---|---|---|
| Forested: | | | | | | |
| | | Site | | | | |
| | Cover Type | G | M | P | Total | |
| | F | 84 | 523 | — | 607 | |
| | L | — | 16 | — | 16 | |
| | Py | — | 32 | — | 32 | |
| | S | 2027 | 5068 | 357 | 7452 | |
| | Pl | 1507 | 5060 | 634 | 7201 | |
| | Total | 3618 | 10699 | 991 | 15308 | 15308 |
| Unforested: | | | | | | |
| | NSR | | | | | |
| | F | — | 68 | — | 68 | |
| | L | — | 15 | — | 15 | |
| | Py | — | — | — | — | |
| | S | 110 | — | — | 110 | |
| | Pl | — | 15 | — | 15 | |
| | Total | 110 | 98 | — | 208 | 208 |
| Other: | | | | | | |
| EPA's | = | 1143 | | | | |
| Other species | = | 446 | | | | |
| Alienated crown | = | 105 | | | | |
| Total | | 1694 | | | | 1694 |
| Total Productive Forest Land | | | | | | 17210 |
| Unproductive for forest: | | | | | | |
| Roads | | 22 | | | | |
| Nonproductive | | 496 | | | | |
| Noncommercial | | 13 | | | | |
| Water (lakes and rivers) | | 45 | | | | |
| Alpine | | 71 | | | | |
| Total | | 647 | | | | 647 |
| Grand Total | | | | | | 17857 |

Editors' note: We interpret the cover types as follows:

F-Douglas Fir (*Pseudotsuga menziesii* (Mirb) Franco); L-Western Larch (*Larix occidentalis* Nutt); Py-*Pinus ponderosa* Douglas; S-Engelmann spruce (*Pice engelmannii,* Parry); Pl-Lodgepole pine (*Pinus contorta,* Dougl).

TABLE 3. Sustained yield projection—area/volume check

(Figures italicised are provided by the user; figures in square brackets are output; see text for further details).

Areas have been reduced by 6.71% for Environmental Protection Area deletions.
Species: Fir. Site: Medium

| Age Class | Area (Hectares) | Volume ($m^3 \times 1000$) |
|---|---|---|
| 1 | 149 | 0 |
| 2 | 9 | 0 |
| 3 | 4 | 0 |
| 4 | 182 | 21 |
| 5 | 30 | 4 |
| 6 | 4 | 1 |
| 7 | 0 | 0 |
| 8 | 110 | 18 |
| 9 | 0 | 0 |

1. *65%* of NSR will be replanted and *35%* will be left to natural regeneration.
2. Delay due to regeneration is *10* years.
3. Delay before replanting is *5* years.
4. Success rate of replanting is *75%*.
5. The current NSR is *63.4372* hectares.
6. The annual loss to roadbuilding is *4%* of harvested area.
7. The first rotation perios is *110* years.
8. Successive rotations are *88* years.
9. Total deduction (due to waste etc) for mature timber is *19.1%*.
10. Total deductions for immature timber is *11.65%*.
11. Improved seed will be available in *20* years.
12. Seed will have a *10%* faster growth rate.
13. Thinning parameters:
    Eldest current stand to be thinned is *40* years.

    Number 1 thinning is at *9* years.
    Number 2 thinning is at *29* years.
    Number 3 thinning is at *46* years.
    Number 4 thinning is at *65* years.

    Stand delpetion is *25%*.

    First thinning to be done on *30%* of the stands.
14. Forest is aged *17* years before calculations begin.

---

15. Current Volume (in $m^3 \times 1000$) is [43].
16. Current Annual Increment is [456] $m^3$.
17. Average Annual Thinning Area is [18.4] hectares.
18. Average Annual Thinning Yield is [288] $m^3$.
19. Sustained Annual Felling Yield is [662] $m^3$.
20. Total Sustained Annual Yield is [951] $m^3$.

---

21. The maximum area to be thinned per annum is *21* hectares.
22. The maximum desired Total Yield per annum is *1200* $m^3$.
23. Summaries printed every *50* years.
24. Calculations cease at year *200*.

TABLE 4. Simulated yields for the forest area shown in Table 3 at 50 year intervals

*Area/Volume Check at Year 50*

| Age Class | Area | Volume | Area Thinned | Thinning Yield | Felling Yield | Total Yield |
|---|---|---|---|---|---|---|
| NSR | 30.3 | | | | | |
| 1 | 86.9 | 0.36 | 63.4 | 0.00 | 0.00 | 0.00 |
| 2 | 89.6 | 2.84 | 238.0 | 2.26 | 0.00 | 2.26 |
| 3 | 73.9 | 3.75 | 0.5 | 0.01 | 0.00 | 0.01 |
| 4 | 97.0 | 9.38 | 0.0 | 0.00 | 0.00 | 0.00 |
| 5 | 58.3 | 6.18 | 0.0 | 0.00 | 0.00 | 0.00 |
| 6 | 5.7 | 0.68 | 0.0 | 0.00 | 0.00 | 0.00 |
| 7 | 100.6 | 12.91 | 0.0 | 0.00 | 13.70 | 13.70 |
| 8 | 0.0 | 0.00 | 0.0 | 0.00 | 12.73 | 12.73 |
| 9 | 0.0 | 0.00 | 0.0 | 0.00 | 3.35 | 3.35 |
| Total | 542.3 | 36.10 | 301.9 | 2.27 | 29.78 | 32.05 |

*Area/Volume Check at Year 100*

| Age Class | Area | Volume | Area Thinned | Thinning Yield | Felling Yield | Total Yield |
|---|---|---|---|---|---|---|
| NSR | 32.7 | | | | | |
| 1 | 87.6 | 0.36 | 65.2 | 0.00 | 0.00 | 0.00 |
| 2 | 86.9 | 2.95 | 217.2 | 2.15 | 0.00 | 2.15 |
| 3 | 86.9 | 5.69 | 267.8 | 4.52 | 0.00 | 4.52 |
| 4 | 86.9 | 8.00 | 185.2 | 4.46 | 0.00 | 4.46 |
| 5 | 120.1 | 13.40 | 0.0 | 0.00 | 0.00 | 0.00 |
| 6 | 22.4 | 2.78 | 0.0 | 0.00 | 11.67 | 11.67 |
| 7 | 9.8 | 1.22 | 0.0 | 0.00 | 6.05 | 6.05 |
| 8 | 0.0 | 0.00 | 0.0 | 0.00 | 11.40 | 11.40 |
| 9 | 0.0 | 0.00 | 0.0 | 0.00 | 0.00 | 0.00 |
| Total | 533.1 | 34.41 | 735.3 | 11.13 | 29.12 | 40.25 |

*Area/Volume Check at Year 150*

| Age Class | Area | Volume | Area Thinned | Thinning Yield | Felling Yield | Total Yield |
|---|---|---|---|---|---|---|
| NSR | 38.8 | | | | | |
| 1 | 115.5 | 0.47 | 77.4 | 0.00 | 0.00 | 0.00 |
| 2 | 109.0 | 3.68 | 233.3 | 2.30 | 0.00 | 2.30 |
| 3 | 91.9 | 5.98 | 219.1 | 3.88 | 0.00 | 3.88 |
| 4 | 86.9 | 8.00 | 217.2 | 5.55 | 0.00 | 5.55 |
| 5 | 86.9 | 10.39 | 0.0 | 0.00 | 0.00 | 0.00 |
| 6 | 6.3 | 0.81 | 0.0 | 0.00 | 31.32 | 31.32 |
| 7 | 0.0 | 0.00 | 0.0 | 0.00 | 4.03 | 4.03 |
| 8 | 0.0 | 0.00 | 0.0 | 0.00 | 0.00 | 0.00 |
| 9 | 0.0 | 0.00 | 0.0 | 0.00 | 0.00 | 0.00 |
| Total | 531.3 | 29.33 | 747.0 | 11.73 | 35.35 | 47.08 |

*Area/Volume Check at Year 200*

| Age Class | Area | Volume | Area Thinned | Thinning Yield | Felling Yield | Total Yield |
|---|---|---|---|---|---|---|
| NSR | 38.8 | | | | | |
| 1 | 111.5 | 0.47 | 83.6 | 0.00 | 0.00 | 0.00 |
| 2 | 111.5 | 3.79 | 278.7 | 2.74 | 0.00 | 2.74 |
| 3 | 111.5 | 7.30 | 272.0 | 4.82 | 0.00 | 4.82 |
| 4 | 111.5 | 10.27 | 250.5 | 6.40 | 0.00 | 6.40 |
| 5 | 46.6 | 5.33 | 0.0 | 0.00 | 29.62 | 29.62 |
| 6 | 0.0 | 0.00 | 0.0 | 0.00 | 4.95 | 4.95 |
| 7 | 0.0 | 0.00 | 0.0 | 0.00 | 0.00 | 0.00 |
| 8 | 0.0 | 0.00 | 0.0 | 0.00 | 0.00 | 0.00 |
| 9 | 0.0 | 0.00 | 0.0 | 0.00 | 0.00 | 0.00 |
| Total | 531.3 | 27.16 | 884.7 | 13.96 | 34.57 | 48.53 |

TABLE 5. Management summary—20 year projection for data of Table 3.

| Year | Area Prepared (ha) | Area Planted (ha) | Cost ($1000) | Area Spaced (ha) | Silva Cost ($1000) | Area Thinned (ha) | Thinning Volume ($m^3$1000) | Felling Volume ($m^3$1000) |
|---|---|---|---|---|---|---|---|---|
| 1 | 0.00 | 0.00 | 0.00 | 0.00 | 0.00 | 2100 | 0.27 | 0.60 |
| 2 | 0.00 | 0.00 | 0.00 | 0.00 | 0.00 | 21.00 | 0.25 | 0.60 |
| 3 | 0.00 | 0.00 | 0.00 | 0.00 | 0.00 | 21.00 | 0.23 | 0.60 |
| 4 | 0.00 | 0.00 | 0.00 | 0.00 | 0.00 | 21.00 | 0.21 | 0.60 |
| 5 | 0.00 | 0.00 | 0.00 | 6.58 | 6.58 | 14.42 | 0.14 | 0.60 |
| 6 | 12.74 | 11.07 | 7.24 | 9.09 | 9.09 | 0.00 | 0.00 | 0.60 |
| 7 | 14.82 | 13.34 | 8.87 | 0.00 | 0.00 | 0.00 | 0.00 | 0.60 |
| 8 | 15.34 | 14.53 | 9.28 | 0.00 | 0.00 | 0.00 | 0.00 | 0.61 |
| 9 | 15.47 | 14.70 | 9.39 | 0.00 | 0.00 | 0.00 | 0.00 | 0.61 |
| 10 | 15.50 | 14.75 | 9.41 | 0.00 | 0.00 | 0.00 | 0.00 | 0.61 |
| 11 | 6.40 | 6.51 | 4.08 | 0.00 | 0.00 | 0.00 | 0.00 | 0.61 |
| 12 | 4.86 | 4.45 | 2.87 | 0.00 | 0.00 | 0.00 | 0.00 | 0.61 |
| 13 | 4.47 | 3.94 | 2.57 | 0.00 | 0.00 | 0.00 | 0.00 | 0.61 |
| 14 | 4.38 | 3.81 | 2.49 | 0.00 | 0.00 | 0.00 | 0.00 | 0.61 |
| 15 | 4.53 | 3.78 | 2.47 | 3.82 | 3.82 | 0.00 | 0.00 | 0.61 |
| 16 | 4.35 | 3.77 | 2.47 | 4.45 | 4.45 | 0.00 | 0.00 | 0.61 |
| 17 | 4.34 | 3.77 | 2.46 | 4.60 | 4.60 | 0.00 | 0.00 | 0.61 |
| 18 | 4.34 | 3.77 | 2.46 | 4.64 | 4.64 | 0.00 | 0.00 | 0.61 |
| 19 | 4.34 | 3.76 | 2.46 | 4.65 | 4.65 | 0.00 | 0.00 | 0.61 |
| 20 | 4.34 | 3.76 | 2.46 | 1.92 | 1.92 | 0.00 | 0.00 | 0.61 |

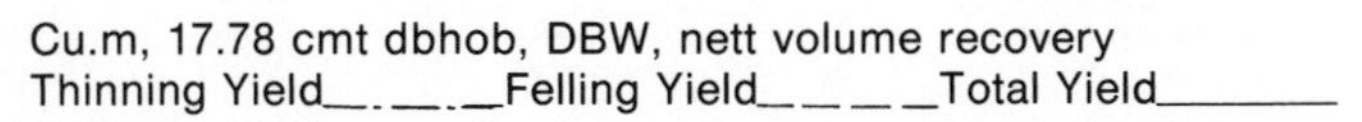


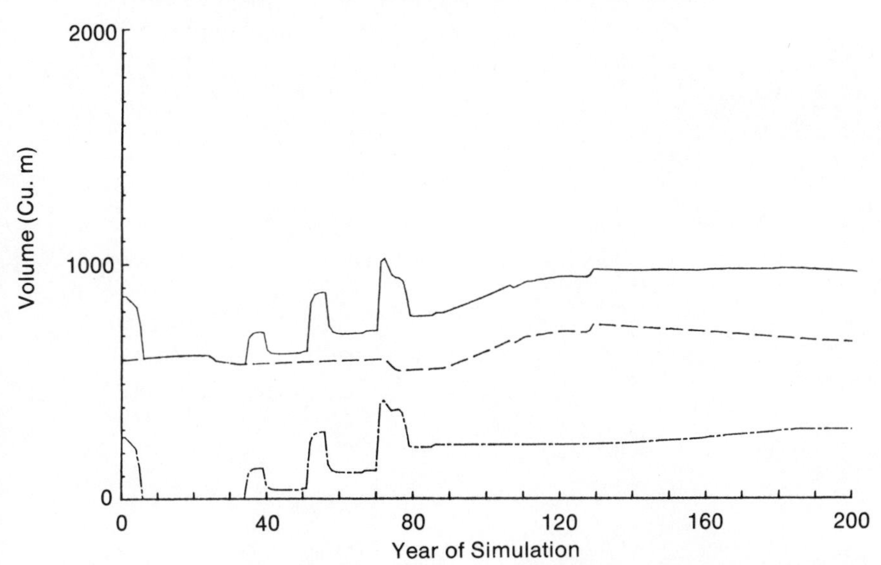


FIGURE 1. Simulation of total yields for data shown in Table 3.

# PART IV

# APPLICATION IN MAPPING AND HARVESTING

# DATA CAPTURE IN THE FIELD

T.J.D. ROLLINSON AND I.D. MOBBS

*Forestry Commission Research Station, Alice Holt Lodge, Farnham, Surrey U.K.*

SUMMARY

A wide range of portable electronic data loggers currently on the market are suitable for use in a forest environment. Microfin data loggers have been selected for recording time-study information, and Epson micro-computers for entering, checking and summarising sample plot measurements. As prices fall and memory sizes increase, wider use of data-capture devices can be expected in both research and management applications.

## INTRODUCTION

Paper, traditionally used for recording forest data, is not the most suitable medium for use in a forest since it is easily damaged and soon gets dirty. In recent years a wide range of portable data loggers and micro-computers have appeared on the market. These machines, developed principally for use in supermarkets for stock control, offer a real alternative to the printed form. In this paper, the use of such machines is discussed for two forest research applications where data are captured in the forest, namely Work Study and Mensuration.

## WORK STUDY

The Work Study Branch of the Forestry Commission's Research and Development Division is responsible for improving operational efficiency and safety. The activities covered include all forestry operations from nursery work and crop establishment through to tree harvesting and timber extraction. There are seven field teams located throughout the country, which carry out method study and work measurement.

In work measurement there is a special requirement to record the time for each component or element of a forest operation. The manual felling operation, for example, is broken into elements such as:

Walk between trees
Chainsaw brash
Fell
Take-down
Sned
Cut-up and put-aside top
Measure and crosscut
etc.

Approximately 20 percent of the total Branch time is spent on time studies. This is primarily for the provision and maintenance of Standard Time Tables and Output Guides, but work measurement also forms an important part of method development.

Traditionally time studies have been carried out using a stopwatch. This has the disadvantage that a considerable amount of time is spent collating and preparing data for analysis, approximately two days for each day of actual study. In many industries such as textile and engineering, where Work Study Officers are dealing with short studies of identical products, this is not important. Studies in forestry are considerably longer and can be up to a full day for a motor manual felling study of 20 trees. Another problem is that, in poor weather, it is difficult to obtain clear undamaged study sheets, despite the availability of waterproof paper.

In order to overcome many of the problems associated with using a stopwatch, electronic data capture was introduced in 1977 with the use of large Rapco data loggers (Rapco Electronics Ltd). The original equipment was built by a local electronics company for the Road Research Laboratory. Similar units were built and modified to incorporate the features necessary for the time study of forest operations. Data were held on magnetic tape in these machines and the tapes posted to Alice Holt for transcription to punched paper tape suitable for computer analysis. This was a cumbersome process but allowed the collection to be carried out automatically, speeding the whole process of data analysis. The introduction of Rapco recorders allowed the development of suitable computer programs for dealing with data recorded on terminals.

Considerable advances in the development of electronic data capture equipment were made between 1979 and 1981, but most of the equipment that became available did not give all the features required for Work Study use. For time-study purposes we wanted a machine with:

a full alpha/numeric keyboard (without a shift key)
no less that 20K of usable memory
a clock, accurate to the nearest centiminute
a secure telephone data-transmission protocol
good operational ranges for temperature and humidity

Then, in 1982, a satisfactory data logger, the Microfin, came on the market (Microfin Systems Ltd.) (see plate 1).

The Microfin is a micro-computer with 8,16,24 or 32K memory which operates under the control of programs written to the specific requirements of each application. The programs may be stored in EPROM or in RAM. It responds interactively to the operator prompting the entry of information and checking the validity of each item before it is accepted and stored in memory. The information is transmitted to the Alice Holt computer by telephone using an acoustic coupler. A secure protocol transmits data in controlled blocks ensuring data integrity. The main power supply is provided by rechargeable nickel-cadmium batteries. Separate batteries protect the memory for up to three months.

Following a six-month evaluation period, the decision was taken to adopt the Microfin as the standard data capture device for Work Study use. Based on these recorders, computer systems have been created so that the Work Study Data Analyst is able to provide a 48 hour return of printouts to the person carrying out the study. A quick turn-round time is important where check studies are carried out for field managers.

Despite rapid developments in new equipment it is intended to continue with Microfin for at least five years to avoid excessive retraining and reprogramming.

**Plate 1. The Microfin datalogger.**

MENSURATION

The Mensuration Section of the Forestry Commission's Research and Development Division manages about 800 sample plots which are individually assessed at intervals of about five years. At each assessment the diameter of each tree is measured together with heights and volumes of a sample of trees. These measurements form the basis of yield studies and are used to derive typical patterns and rates of tree growth on a range of sites. Some sample plots were set up to determine the effects of planting trees at different spacings and to investigate the response of trees to different thinning and respacing treatments.

Until the early 1970's all data were collected and processed manually. Thus foresters visited sample plots, measured the trees, recorded the measurements on paper, and then, back in the office, summarised the data and worked out the results. If any errors were detected the forester would return to the plot to check and, if necessary, correct the measurements. The paper records were then sent to Alice Holt for storing. This procedure had a number of disadvantages. Foresters used to spend approximately half their time calculating results and made frequent errors. The paper forms were often damaged in the field and were difficult to read after many years of storage.

More recently the introduction of computers has allowed foresters to send their field measurement forms direct to Alice Holt where the time-consuming calculations were performed by computer. This procedure, however, still had its disadvantages. Although the calculations were carried out correctly, errors were sometimes introduced when data were copied from the field forms onto the computer. Also, delays due to the postal service and data preparation meant that results were received

Plate 2. The Epson HX-20 microcomputer.

by the forester long after he had left the sample plot, so that he was unable to check the data easily.

In recent years, the availability of portable computers has offered the prospect of radically changing these procedures. The ability to process results in the forest means that postal delays and copying errors can be eliminated. The specification for a suitable machine included the following requirements:

a minimum of 16K usable memory
a large, liquid crystal display
an easy-to-use keyboard
reasonable weather-proofing
and, most importantly, programmable by staff at Alice Holt.

The machine that is being used by the Mensuration Section to collect data in sample plots is an Epson HX-20 micro-computer (Epson (UK) Ltd) (see plate 2). It is programmable in BASIC and comes with standard 16K RAM memory, a 20 character, 4 line liquid crystal display, a 24 column dot matrix printer, a built-in time and calendar clock, a full typewriter keyboard, a speaker for audible warnings, an RS232 interface, and as an option a microcassette drive. In its application for data collection in forestry experiments the memory has been expanded to 32K. The Epson weighs approximately 1.7Kg and measures 29.0 × 21.5 × 4.4cm and is powered by nickel-cadmium rechargeable batteries with a 40 hour capacity. These various features of the machine offer a great flexibility when it is used in the field. The built-in printer allows the forester measuring the tree to produce a hard copy of all data as they are entered, and to print intermediate results. The micro-cassette provides another medium for data storage. The transfer of information to and from the main sample plot database is achieved by connecting the Epson to an acoustic coupler and transmitting the data via a standard British Telecom telephone through a modem at Alice Holt. This translates the signal into characters for input to the main computer.

In practical use the flow of information is as follows:

*Collection of previous measurements.* In the office, the field forester dials the Alice Holt computer. The previous measurements for the sample plot are transmitted over the telephone line to the Epson HX-20. Alternatively the data are recorded on cassette at Alice Holt and the micro-cassette is sent in the post to the field forester who loads it into the Epson.

*Field recording.* At the plot, the forester collects the new measurements in the sample plot. These are checked against the previous measurements as they are entered. If an apparent error is detected such as a diameter being smaller that its previous measurement, the machine emits a bleep and displays a message on the screen asking the forester to check the measurements. The program calculates intermediate results, such as tree volumes, as data are entered.

*Field processing of data.* The computer checks that all the measurements have been entered and processes the data. Any errors are displayed on the screen so that additional measurements or re-measurements can be done while the forester is still in the plot.

*Transmission of data to the main computer.* Back in the office, the forester transmits the data over the telephone line to the Alice Holt computer. If he experiences difficulty with transmission, he sends a micro-cassette containing the data in the post to Alice Holt where it is read directly on to the computer.

Use of the Epson has a number of advantages over previous systems. Firstly, the machine can be programmed by staff at Alice Holt so that programs for data collection can be tailored to the users' requirements, and can be easily developed and maintained. Secondly, as the Epson is a computer, preliminary processing of data can be done in the field so increasing job satisfaction, greatly reducing delays in sending data to and from Alice Holt, and allowing field teams to check and amend data before transmission for permanent storage. Thirdly, the built-in printer allows the forester to keep a permanent record of measurements and results. This is important during the early stages of introducing computerised data collection to staff who are used to working with paper forms. Fourthly, the computer power allows the forester to manipulate data, for example, to sort tree diameters into diameter classes. Finally, data can only be entered in a particular format and in a particular order, which ensures that data are consistent between measurements. The program prompts the forester for information, and can check that all trees have been measured.

### DISCUSSION

Complete with an acoustic coupler, a 24K Microfin terminal costs about £850 and an Epson micro-computer a little over £700. The total annual cost of each machine including maintenance is unlikely to exceed £300. With agency rates for basic grade Foresters in the region of £100 per day, it is not difficult to justify the purchase of data collection devices when time savings can so easily be demonstrated.

With fast moving technology, new machines are continually appearing on the market. At some stage a choice has to be made. The Microfin and Epson terminals described in this paper fulfil the requirements of the Work Study and Field Survey Branches of the Research and Development Division. Having created computer systems to deal with each type of data terminal, neither machine is likely to be replaced in less that 3 to 5 years.

Costs of portable data-capture devices are falling at the same time as processing capabilities and memory sizes are rising. It is expected that such machines will be used increasingly for data collection and manipulation in both research and management applications. Wherever large quantities of data are collected on a regular basis, the use of these machines should be considered. A look into the future might reveal the following field applications:

*Sawlog measurement.* The forester enters measurements of length and top or mid-diameter of sawlogs. The data capture device calculates the volume of the individual sawlogs and summary totals.

*Tariffing.* The forester enters measurements of breast height diameter, length and mid-diameter of volume sample trees, the breast height diameters of girth sample trees and a count of the total number of trees. The data capture device calculates and prints the total number of trees, the estimated volume (with confidence limits) and any other details that are required such as mean diameter and mean volume of the stand.

*Forest inventory.* The forester enters survey data such as forest name, grid reference, compartment number, species, age and yield class, and any other relevant information such as stocking, windthrow hazard classification and terrain class. The data capture device stores the data for transmission to a central computer where the data are analysed and permanently stored. Portable data capture devices have been used by the Forestry Commission's Field Surveys Branch to recover and transmit the ground survey data for the recent census of woodlands and trees.

*Stock control in sawmills.* The sawmill manager enters details of the movements of timber during its progress through the sawmill and warehouses. The data capture devices provide up-to-date information about the location and amounts of different kinds of timber in stock, so facilitating stock control, production and despatch operations.

# DIGITAL MAPPING, TOPOGRAPHIC MODELLING AND GEOGRAPHICAL INFORMATION SYSTEMS IN FORESTRY

A.J. CROSBIE

*Dept. of Geography, University of Edinburgh EH8 9XP*

SUMMARY

The use of data bases in forestry management has grown over the past twenty years. Digital mapping permits visual display of this information and, when combined with topographic modelling, provides an assessment of options and an appraisal of costs in working forests. Together these are powerful tools in planning and implementing production policies and they have become increasingly so with the development of micro-computers and the availability of relational data bases. The ability to overlay complex information from a relational data base and present this data in graphical or map form is the central feature of geographical information systems (GIS). Digital mapping is also significant in monitoring changes either through the use of remote sensing or recurrent censuses. Remote sensing from satellites such as Landsat or Spot is unlikely to be of direct benefit to day to day forest operations in the United Kingdom where the mosaic of woodlands requires very high resolution. However, the increasing benefits of multi-spectral scanning from aircraft will overcome this problem.

INTRODUCTION

The development of information technology is of direct benefit to forestry and forest management. The need for resource data for inventories, silviculture, planning, production control and policy making has long been apparent but it has also become necessary to integrate it with wider land use data. Forests exist in Britain as part of the landscape and they both influence, and are influenced by, the land around them. The relationship between woodland and agriculture is obvious but with the increasing emphasis on recreation and conservation there is a requirement for information on distance from towns, access, and the distribution of population. The use of maps to meet these demands in forestry is common practice but the development of automated cartography and relational data bases has created tools of great significance for decision making.

Maps, however generated, are only as good as the data from which they are compiled. In the past, data have often tended to be discrete and uneven in their form and collection due to practical limitations and varied objectives. Statistical data, by their very volume, have to be compiled into tables, lists of figures, graphs and histograms for convenience. Even so, no one can comprehend or easily utilise long lists of tables and, for this reason, census data—be they of people or forests—have been presented in map form for the past fifty years. Now, methods of collecting data have improved with much being in machine readable form, although it is worth emphasising that for field information ground checking remains essential. The use of

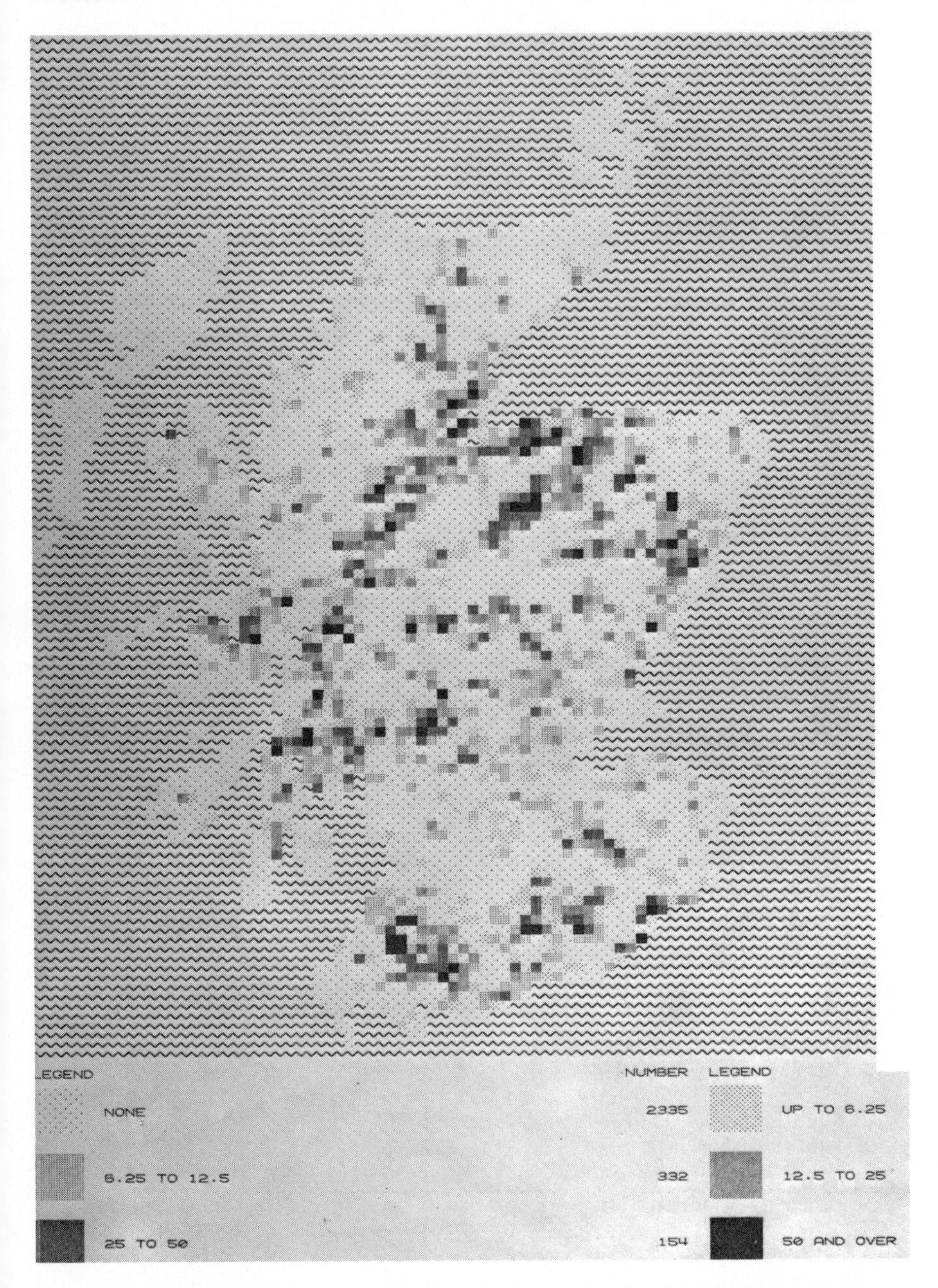


FIGURE 1. A grid square map of Scotland indicating the percentage of each 5km square occupied by woodland.

such data, however, depends on the speed of manipulation and analysis and this has been dramatically altered by the use of computers.

However, the ability to relate such data to specific locations is not automatic. Correlation with boundaries or natural features is not universal and usually a map has to be specially drawn either by hand or computer. Where two sets of information require to be linked either a new map must be compiled using the joint information or an overlay is required. The compilation of vast data banks combined with the development of digital mapping and computer graphics has not only revolutionised map making but also altered the emphasis from the map to the information. Traditionally, the constraints of time and technique meant that the type of information had to be carefully selected for presentation in map form for analysis. Now, with sophisticated skills in information management, maps will be generated, cheaply and rapidly, to the unique requirements of the individual or organisation. All data in machine readable form will be available for analysis and correlation, but equally the limitations of the data will become the main constraint. For example, the geographical location, distribution and size of woodlands in Great Britain may be mapped but other features such as age, rate of loss or gain in planting, or ownership may be of equal significance. The three key factors in data analysis remain availability in terms of accuracy, and comparability.

There are two basic methods of digital mapping. These are:

1) through the use of grid squares related to a national or a consistent geographical grid;
2) through digitising map polygons into units of varying size which may be mapped in relation to digitised boundaries of administrative or geographic regions.

The first method is valuable in providing cheap storage of data and ease of processing of data of a generalised nature. It is essentially an averaging system but is very useful in presenting information over extensive areas, such as in Canada (Bonner, 1982), where the demonstration of gradation is required and the administrative divisions are on a similar base. A land use map of England and Wales has been constructed using this technique but it is not an effective visual presentation for local areas and it does not permit details or precise comparisons. For these reasons it is not really appropriate for forestry work in this country where the units are often small, irregular and part of a landscape mosaic (Figure 1).

The second method assigns digital co-ordinates, related to the National Grid, for all data to lines and points on a map which are then stored in a computer (Figure 2). The advantages of this method lie in the detail possible in map production and the wider range of analytical work that can be carried out. In terms of precision and high quality production, this method is outstanding.

Both methods offer a range of possibilities. They permit scale changes by simple computer manipulation; they can link nominal and ordinal data; they allow constant revision of both data and boundaries; and they provide search routines by grid squares, polygons, points in polygons and by pathways.

## TOPOGRAPHIC MODELLING

Apart from producing maps of resource data, it is also possible to create models or cross-sections of topography. Information about relief, drainage and other physical or economic features are stored in memory and then produced in the form of a three

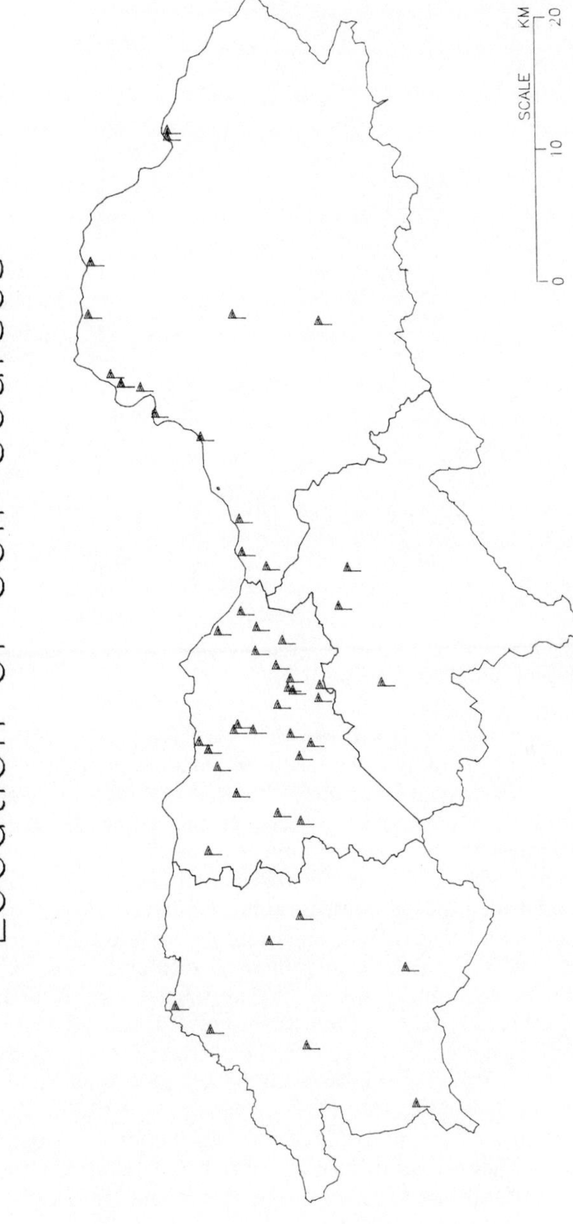


FIGURE 2. The location of golf courses in Lothian using digitised map polygons to give point locations and administrative boundaries.

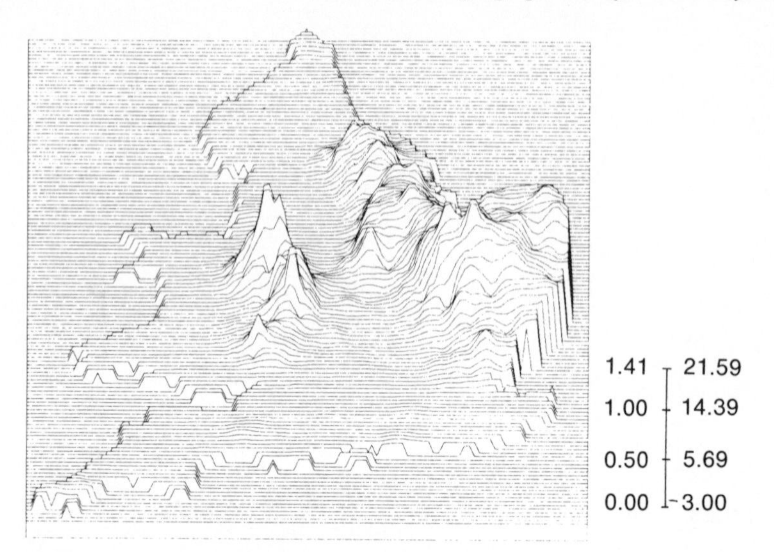


FIGURE 3. A computer model to illustrate by vertical exaggeration the significance of sugar beet cultivation in England and Wales (1973).

dimensional perspective drawing known as a block diagram (Figure 3). The value of this form of mapping to forestry is varied and significant. For example, in the construction of a working plan for an area of forest it is possible to illustrate the yield by location within the forest. Alternatively, in calculating the value of timber produced from a forest, the extraction costs vary with each working section. This can be related to the economic costs involved in extraction, such as in road construction, taking into account the need for cuttings, gradients and so on.

The use of models and digitised mapping packages will also allow consideration of other strategies in extraction by relating these to surrounding conditions e.g. the provision of public transportation routes. Similar practices may be used in planting strategies where the use of packages (such as PREVIEW) will permit a visual display of the changing landscape, from various directions, as trees grow up. Equally, the effect of guard rows or the impact of thinning may be represented by such packages.

These simulation techniques provide both a means of testing the validity of predictions about growth conditions and of assessing the advantages and limitations of possible options. They are widely used in research and the work of the Institute of Terrestrial Ecology should be mentioned here (Bunce and Heal, 1984).

## GEOGRAPHICAL INFORMATION SYSTEMS

The integration of digital mapping and relational data base systems into geographical information systems has further extended the ability to manipulate the two basic units of space and time in development applications. An experiment carried out in the

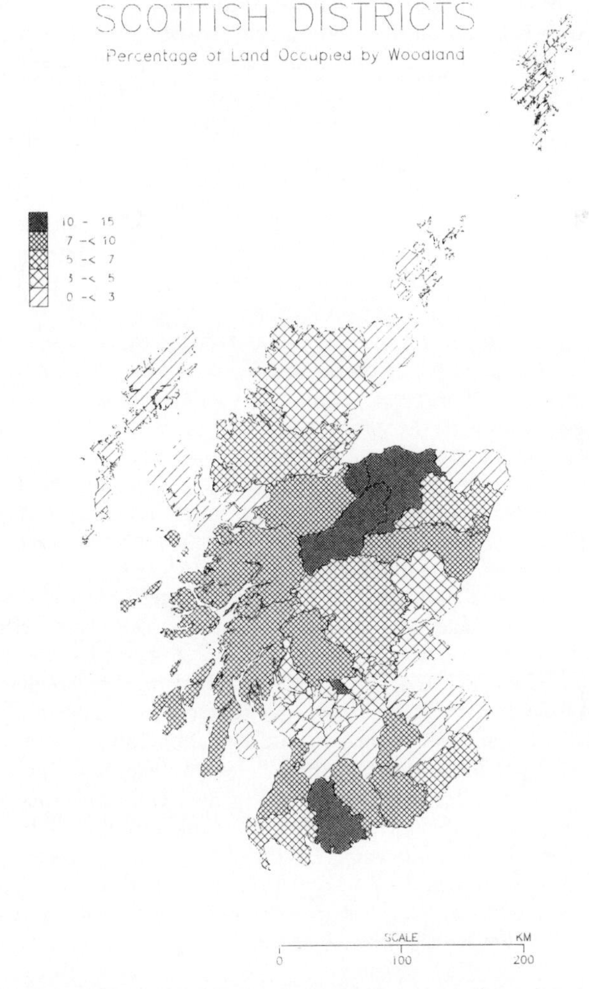


FIGURE 4. A map generated by combining TRIP and GIMMS to indicate the percentage of land occupied by woodland in Scotland (compare Figure 1).

Planning and Data Management Service of the Department of Geography, University of Edinburgh, using a Rural Land Use Information System (RLUIS) on a section of Fife illustrated the potential of this method. All information on this specific geographical area relating to planning controls, forestry or agricultural development was brought together in order to demonstrate the effect of change in any one area or form of land use on the others. Two packages were used in this experiment: the Tourist and Recreation Information Package (TRIP) and the Geographical Information Manipulation and Mapping System (GIMMS). The former incorporates a grid square framework for the whole of Scotland at the 5 kilometre and 1 kilometre level. Figure 4 gives an example of the type of map that could be generated by combining these two packages. Data sets were compiled in three forms—linear (or boundaries), point by six figure grid reference, and areal by reference to a unique grid square. The latter is based on digitised polygons which gives a greater degree of precision and is particularly useful for mapping line and point data such as roads, railways and pipelines. By interfacing the two systems, it was possible to generate maps in great detail and to undertake a wide range of analytical work on the spatial data.

### MONITORING CHANGE

Digital mapping is also based on data derived from remote sensing. In the past, conventional aerial photography has proved valuable in complementing planimetric maps and ground surveys for forest planning and production schedules but it is slow and costly to collect. Much depends on the scale: for country wide surveys they are indispensable in recording forest conditions at a given time and place, while at a more local scale they are useful in calculating total forest area, volume and growth, the kind, age and condition of the trees, general accessibility and the location of timber in relation to transportation. At the same time, monitoring landscape change is essential in any policy of economic or responsible land use, and this requires repeated surveys at frequent or regular intervals. The rate and direction of change is assessed by measuring the physical and biological characteristics of the system as it develops and this provides a framework for measuring responses to particular courses of action.

Satellite imagery using multi-spectral scanners has given very useful and colourful presentations of the landscape and its component parts. This is most effective in the wide, extended forest covers of North America or Eurasia but the resolution is inadequate for the small, varied mosaics with which we are familiar in Britain. Advances in this field are rapid, however, and the resolution is constantly improving although not to the degree used in military satellites. Both Landsat 5, launched on 1st March, 1984, and Spot, the forthcoming French satellite being launched by European Space Agency, have much improved capabilities. The thematic mapper in the former gives a pixel size of 30 by 30 metres, while Spot provides pixels at 20 by 20 in colour and 10 by 10 metres in black and white. Even at this resolution, however, they fail to meet the requirements of scale for practical forestry. More recently, the use of multi-scanners from conventional aircraft demonstrates that these will prove more useful for detailed work in the British forestry scene.

### CONCLUSION

Forestry is the growing of trees including commercial timber management and conservation and it is a skill requiring considerable knowledge and information. Information technology does not in any way replace the professional ability of the forester but it does provide an extremely powerful tool for data analysis which permits more balanced and informed courses of action.

### REFERENCES

Bonnor, G.M. (1982) *Canada's forest inventory, 1981* Forestry Statistics and Systems Branch, Canadian Forestry Service, Department of Environment.

Bunce, R.G.H. and Heal, O.W. (1984) Landscape evaluation and the impact of changing land-use on the rural environment: the problem and an approach. *in* Roberts, R.D. and Roberts, T.M. (eds) *Planning and ecology*. London. Chapman and Hall.

# SOME APPLICATIONS OF THE HIGHLAND REGION RURAL LAND USE INFORMATION SYSTEM TO STUDIES OF FORESTRY AT A STRATEGIC LEVEL*

R.G.H. BUNCE[1], C.J. CLARIDGE[2], C.J. BARR[1], M. BALDWIN[2] AND R. CAMERON[2]

[1]*Institute of Terrestrial Ecology, Merlewood Research Station, Grange-over-Sands, Cumbria LA11 6JU.*
[2]*Highland Regional Council, Department of Planning, Glenurquhart Road, Inverness.*

SUMMARY

The establishment of a database to aid rural planning by the Planning Department of the Highland Regional Council is described. When used in conjunction with a model, the data base can be employed to examine potential areas of conflict between different land uses. Other possible uses of the system such as forecasting levels of timber production and consequental requirement for road construction are outlined.

## INTRODUCTION

Highland Region is some 25 500km$^2$ in extent and covers 33 per cent of the area of Scotland and 10 per cent of Great Britain. Much of the Region is remote and the small population density (under 7 persons per 1km$^2$ compared with 229 for the UK as a whole) emphasises the importance of the rural aspects of planning. The Region has many highly valued wilderness areas designated both for wildlife conservation and for scenic importance. On the other hand, the pressures upon the land particularly for forestry have greatly increased in recent years, emphasising the need for strategic studies.

The Planning Department of the Regional Council has the responsibility for, among other issues, the making of strategic plans involving rural activities and therefore needs to identify key land use changes. Although there was extensive fragmented information on the rural environment, a co-ordinated view was not available. Because of the large area of the Region and limited staff, a sampling system developed by the Institute of Terrestrial Ecology was adopted to build up an appropriate data base (Bunce *et al* 1981; and Bunce and Heal 1984). This procedure classifies the land surface into environmental strata which may then be use to sample a range of ecologically related factors. In Highland Region an information data base was set up using only physical data for each one kilometre square drawn from maps for all the squares within the Region. One example, for the distribution of woodland, is shown in Figure 1.

These data have then been used to classify the squares into 24 environmental strata, termed land classes. A sample of 8 squares per class were surveyed by using a

*This paper was submitted as a poster presentation at the Conference.

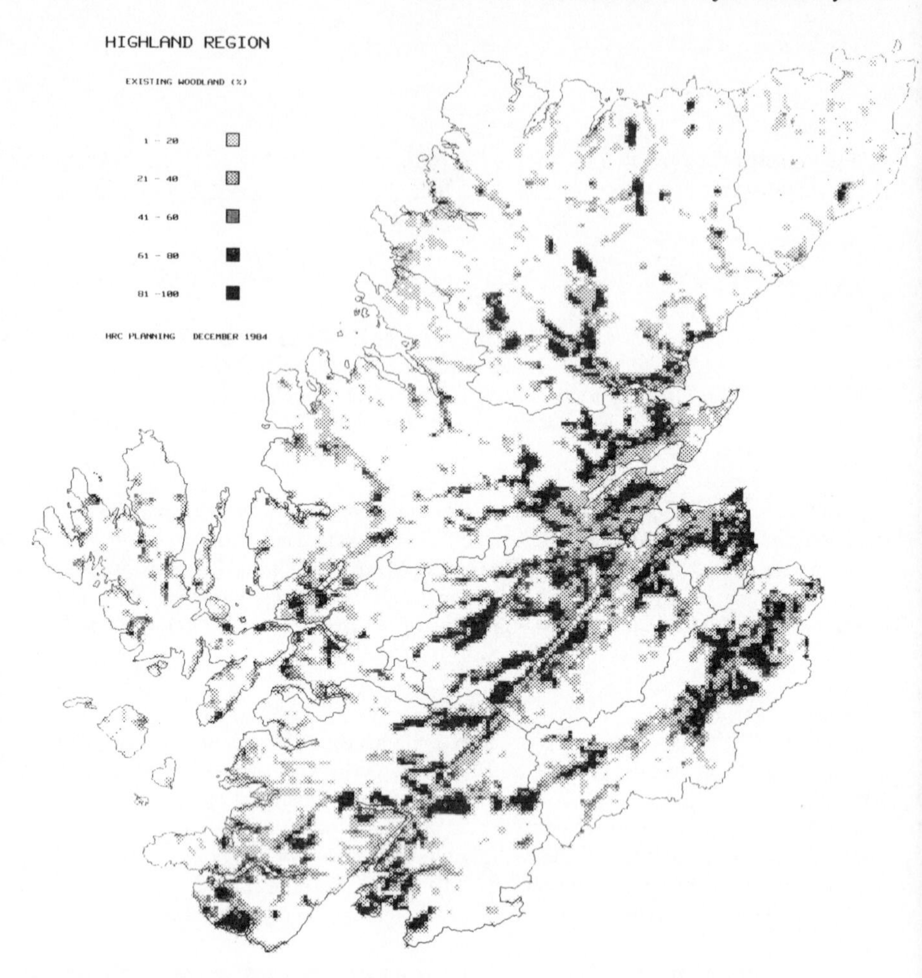


FIGURE 1. Presence of woodland in all 1km squares in Highland Region.

field assessment booklet and 1:10 000 maps for recording (i) topography, (ii) natural vegetation and agriculture, (iii) recreation and boundaries, (iv) woodland and (v) structures and roads.

A fundamental feature of the approach has been to use a micro-computer (Baldwin 1984), located in the Planning Department, to set up the data base and analysis procedures. This represents a cheap and flexible alternative to extensive centralised systems and has enabled the Planning Department to develop and control the data base.

The data from the sample squares were used in the manner described by Mitchell *et al.* (1983) to enable economic comparisons to be made between agriculture and potential forestry on defined units of land. A model is developed, based on the sample data, which is able to provide estimates of the technical potential for forestry before

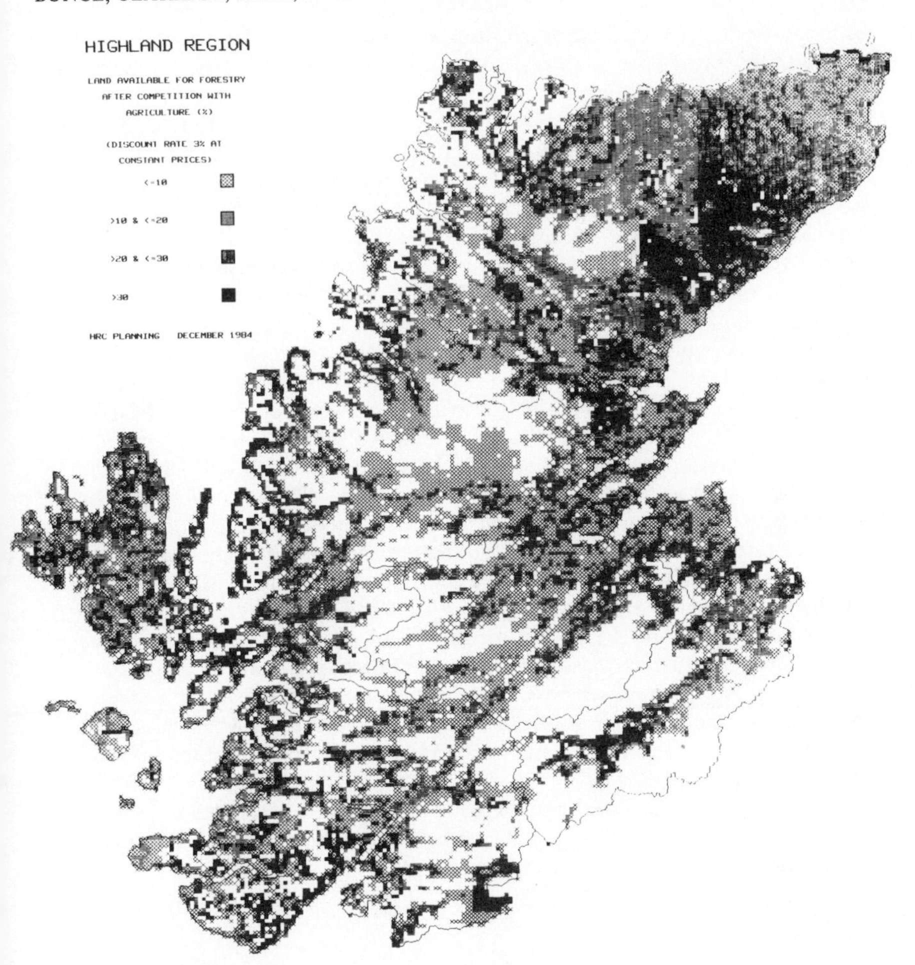


FIGURE 2. Prediction of land available for forestry after competition with agriculture (%) in all 1km squares of Highland Region.

constraints such as 'areas of scientific importance' are applied according to their occurrence on given land units. In this way the Planning Department can examine the effects of their own policies and then progressively identify potential conflicts between the interests of various agencies.

Already the overlap between the designated sites of the Nature Conservancy Council and the highly-valued areas for red deer have been compared with potential forestry. Adjustments can also be made to various economic parameters, such as discount rate and timber price, in order to assess the sensitivity of the system. Mean values are calculated for each land class and then used for prediction over the Region as a whole. For any one kilometre square account is also taken of certain technically unplantable land, such as open water or existing forest, before the estimate is made. Figure 2 shows a preliminary example of the prediction of potential forest. The areas with no forestry predicted are either due on the one hand to exposure on coasts or high altitude preventing planting, or on the other due to forestry being less competitive than agriculture. Different patterns can be compared with alternative forest scenarios, in order to examine the stability of the prediction. The total predicted area of the woodland is comparable to that available from independent sources indicating the validity of the sample. Predicting the land suitable for forestry is only one area that the model can be used for, since production levels of timber could also be forecast, as well as job opportunities. Currently, the Regional Council are examining various sections of the model in order to improve the sophistication of the predictions, since at the moment generalised assumptions are made for economic factors, such as the construction of roads, that can in fact be made specific to individual squares. Such progressive development to the data base demonstrates the flexibility of the system.

In 1984 the Regional Council decided to undertake a survey of woodland using the same series of samples, as they were concerned with the loss of broadleaved woodland, the lack of planting of broadleaves, and the lack of a systematic study of the overall woodland resource, including scattered trees. The objectives of the survey were to establish the economic, ecological and landscape significance of woodland in the Region, to assess its state and the factors affecting its management. Likely changes also needed to be identified as was the scope for action to ensure continuity of woodland. The data are currently being analysed, but in the present context demonstrate the use of the system for rapid and effective survey, as preliminary figures show close similarity with comparable Forestry Commission statistics.

Various other data bases have been incorporated such as peat reserves, the distribution of gaelic speakers and the census data. Other extensions planned are the land capacity for agriculture classes of the Macaulay Institute, figures for acid rain and land tenure.

The system will overcome many of the limitations associated with the traditional plan-making processes and will lead to a better understanding of planning and its environment.

REFERENCES

Baldwin, M.B. The use of a micro-computer for mapping purposes, Highland Region Information Notes No. 2. Highland Region, Inverness.

Bunce, R.G.H. and Heal, O.W. (1984) Landscape evaluation and the impact of changing land-use on the rural environment: the problem and an approach. In: Planning and ecology, edited by R.D. Roberts and T.M. Roberts, 164-188. London: Chapman and Hall.

Bunce, R.G.H.; Barr, C.J. and Whittaker, H.A. (1981) Land classes in Great Britain: preliminary descriptions for users of the Merlewood method of land classification. Merlewood research and development paper No. 86. Grange-over-Sands: Institute of Terrestrial Ecology.

Bunce, R.G.H.; Claridge, C.J.; Barr, C.J. and Baldwin, M.B. (1984) The use of simple data in the production of strategic sampling systems—its application to the Highland Region, Scotland. In: Methodology in landscape ecological research and planning. Vol. 4.: Methodology of evaluation/synthesis of data in landscape ecology, edited by J. Brandt and P. Agger, 167-171. (1st int. Seminar of the International Association of Landscape Ecology). Roskilde: Roskilde University Centre.

Mitchell, C.P.; Brandon, O.H.; Bunce, R.G.H.; Barr, C.J.; Tranter, R.B.; Downing, P.; Pearce, M.L. and Whittaker, H.A. (1983) Land availability for production of wood energy in Great Britain. In: Energy from biomass, edited by A. Strub, P. Chartier and G. Schleser, 159-163. (2nd E.C. Conference, Berlin, 1982). London Applied Science.

# SIMULATION TECHNIQUES AND HARVESTING SYSTEMS

I.D. MOBBS

*Forestry Commission Research Station*
*Alice Holt Lodge, Farnham, Surrey U.K.*

SUMMARY

When expensive machines are required to work together in complex harvesting systems, it is useful to be able to isolate and examine those factors that affect harvesting costs. This is difficult to do in practice, but is possible with computer simulation.

Cable-crane extraction is used as an example to illustrate the type of questions that can be answered by simulation. Some general results emerge but many answers relate only to specific site and crop conditions.

## INTRODUCTION

Foresters in a number of countries have used computer simulation for operational planning and control purposes. A number of papers describing the application areas were published as proceedings of the IUFRO Workshop/Symposium held at Wageningen (Anon 1978).

British foresters are practical men who do not really turn to computer packages for help in deciding how to use harvesting machines. Some have, however, been happy to accept standard items derived from mathematical models as a basis for negotiating piece work sales and for forecasting production.

As harvesting systems become more complex and machines more expensive, it is not so easy to experiment for the purpose of minimising cost. Bringing machines together on to one site to deal with specific crop conditions is usually expensive and sometimes impossible. In these circumstances computer simulation can answer a number of questions, some of a general nature, others site-specific, which can make the difference between profit and loss on a harvesting operation.

## MACHINE PRODUCTIVITY

Most harvesting machines are easily moved and can be separated from others in a harvesting system by time or produce buffers. Where this is possible there is little need for any detailed examination of performance by any means other than the on-the-spot testing of new ideas.

Figure 1 illustrates the concept of machine productivity in relation to material awaiting processing.

Often the conditions in which a machine works are stable and fall well within the safe zone on the right of the figure. Over a week's production may separate one machine from adjacent ones in the system and only lengthy breakdowns in preceding operations will significantly affect output. The produce buffer is sufficient to protect the machine from any problems affecting its neighbours in the system.

Within the critical zone (on the left of the figure) small gains in output could be

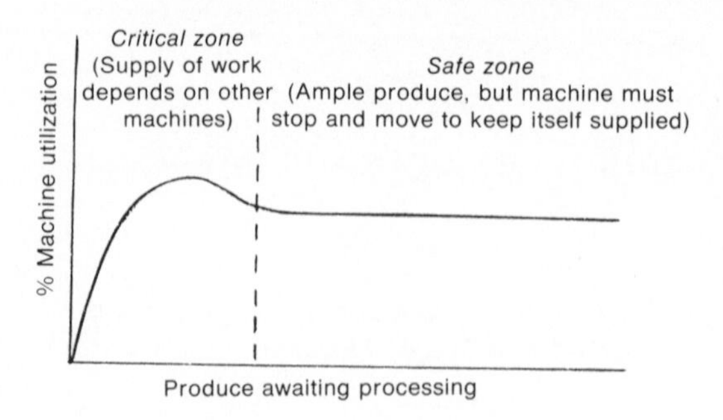


FIGURE 1. Machine efficiency

achieved if material were fed to the machine and produce removed as required, to leave the working area uncluttered. There would be no need to move the machine. The gains, however, could well be offset by losses elsewhere in the system.

Working conditions for machines brought together to work in a close system are likely to be in the critical zone. Any problem with one operation can affect the next. Machines may have to wait for work and freedom of movement may be restricted by the activities of neighbours. It is when individual machine productivity is constrained by the system of working imposed, that computer simulation can help to direct attention to the operational factors that most affect overall harvesting crops.

### CABLE-CRANE PLANNING

The cable-crane is one example of a machine which not only depends on another to keep the quantity of produce within prescribed safety limits at the landing, but is also expensive to move. Figure 2 illustrates the major component of a typical cable-crane system. The carriage runs in a fixed skyline which is anchored at each end, movement of the carriage being controlled by the haul-in and haul-back ropes. When the carriage is correctly positioned, the haul-in rope is released to allow the chokerman to hitch the end to the nearest load of timber.

Conventionally extraction racks are 40m apart and parallel, so that the maximum side-haul distance is no more than 20m. Side-haul distance is here defined as the distance the rope is taken away from the carriage to reach logs within the extraction strip.

The time to set up and take down a cable-crane system to serve one rackway is measured in terms of hours rather than minutes. Total set-up and take-down for a particular area is related directly to the number of times the job is done (Figure 3). Side-haul distances, on the other hand, are high when there are few extraction routes and low when there are many. Load-time in relation to side-haul distance is likely to rise significantly as 'practical' side haul distances are exceeded.

A further feature of cable-crane operation is that a secondary extraction tractor, either a skidder or forwarder, is often required on the road to clear produce from the cable-crane when heaps become too large. In cable-crane country stacking space is often limited and the cable-crane is unable to continue extraction without some clearance of the accumulated produce.

In planning cable-crane extraction a large number of questions can arise, among them:

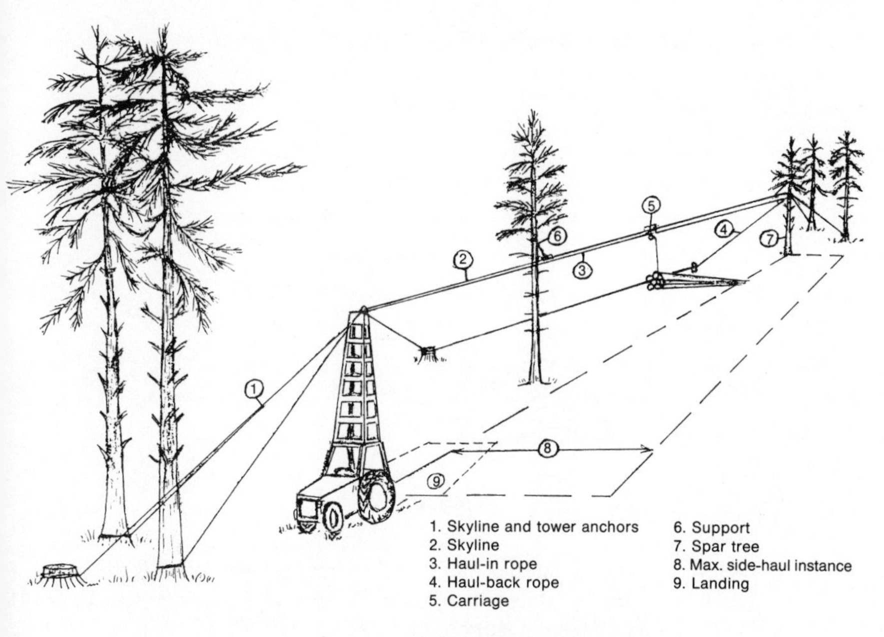


FIGURE 2. Components of a cable-crane system

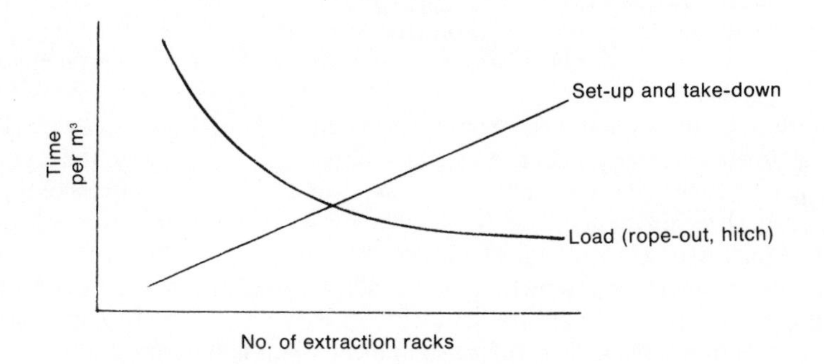


FIGURE 3. Effects of rackway spacing on extraction time

1) Is the conventional maximum side-haul distance appropriate in all circumstances? Is this distance a critical factor in cable-crane extraction costs?
2) Can secondary extraction be avoided by reducing side-haul distance? If so, is this a cheaper option?
3) If a skidder or forwarder is needed for secondary extraction, what proportion of its time is available for other extraction work in the immediate vicinity? Can it be spared for long enough to do useful work elsewhere?
4) What is the best way to extract from a site which needs cable-crane for part of the area only? Should the whole site be cleared by cable-crane or should the tractor-accessible part be cleared by skidder or forwarder?

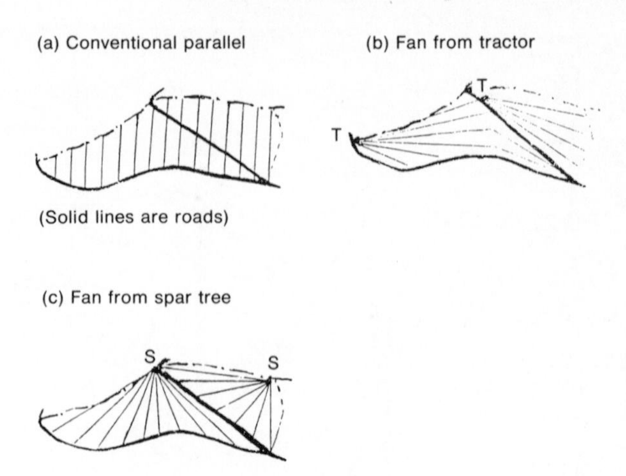


FIGURE 4.

5) Is the conventional parallel pattern of extraction best? What is the effect of using fan patterns (a) from the tractor (b) from the spar tree? (See figure 4)
6) How would engineering work to increase the size of landings and so reduce handling of extracted material, affect overall extraction costs?
7) How does load size affect productivity of the cable-crane and the intervals between visits from the secondary extractor?
8) How would produce presentation resulting from contour-felling affect cable-crane costs?

The answers to these questions are not immediately obvious. If two areas are cleared using different extraction patterns, are the differences in costs due to the extraction pattern used, to the different crop/site/landing conditions or to the effects of the New Year celebrations on the operator. How is it possible to keep other factors constant so that one only can be investigated?

Computer simulation is a method that can be used to give comparative costs. By simulation an area can be 'cleared' as many times as we would wish using different extraction patterns, secondly handling equipment and assuming various crop and site characteristics.

## THE COMPUTER PROGRAM

The simulation program which has been written uses time element equations derived from the Forestry Commission Work Study database.

The program simulates the actual extraction process using four phases in the cable-crane time cycle, adding timber to the roadside stack at the end of the fourth phase. The phases are:

Move empty carriage into the wood;
Pull out the rope and hitch to a load of timber;
Haul the load to the road-side;
Unload.

A simplified flow diagram of the program is given in the Appendix.

The extraction may be carried out using any one of the three basic patterns (Figure

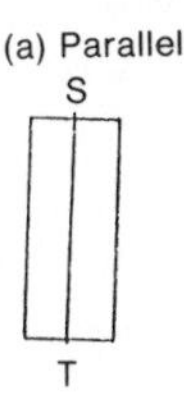

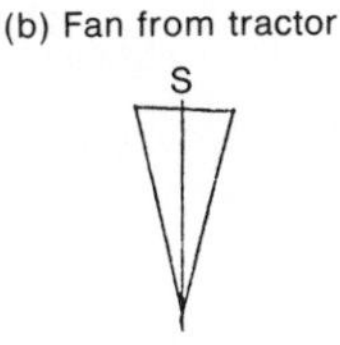

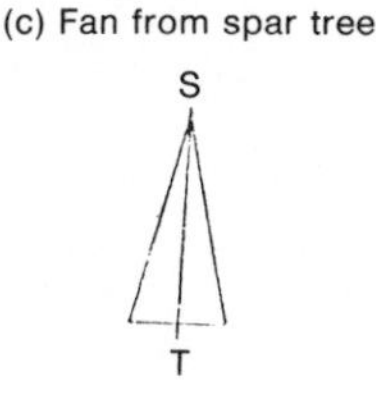


FIGURE 5.

5), taking account of any patches of ground without produce within, or at either end of the rackway.

Individual load distances from the cable-crane have been obtained by random sampling from a uniform distribution bounded by the current side-haul limit. The limit is constant for conventional parallel extraction patterns, but proportional to distance from the tractor for fan systems.

The user is asked to specify which secondary extraction tractor is available. It must, of course, be one for which cyclic times for road work are known. When a large machine, such as the Falstone tractor, is used the program assumes that it arrives on site just before the volume at the road-side exceeds the safe capacity and leaves when the landing has been cleared, or when sufficient capacity has been created to accept all produce still remaining in the rackway.

Smaller machines (such as the Thetford Hydratongs) start work when the cable-crane has extracted 10 cubic metres. They continue on site until the rackway is cleared, because the machine extraction speed roughly matches that of the crane.

Loading is allowed to take place only when the cable-crane carriage is in the wood. If the tractor arrives in any other part of the cable-crane time cycle, it has to wait. Waiting time is accumulated and printed as a percent of total secondary extraction time.

Travelling time to and from the job is assumed to be half an hour each way when the tractor has other work in the vicinity. If the machine does not leave the site, only the machine cost is included for the full period. When the tractor has no work to do, the operator is assumed to be productively employed in crosscutting or stacking activities chargeable to a different account head.

## SOME EARLY RESULTS

To prove that the program worked, times obtained were compared with standard time tables. There were no differences that could not be accounted for by rounding errors.

The general results set out below emerged from simulating skyline pole-length extractions with an Igland cable-crane and a Falstone skidder. It was soon apparent however that cable-crane costs are site-specific and depend on detailed site and crop information.

1. Side-haul distances and produce density (Figure 6). The side-haul limit has a considerable effect on extraction costs when produce density is low. At normal clear-felling densities side-haul distance is not critical to the cable-crane cost, but affects the number of visits required from the skidder.
2. Side-haul distance and secondary extraction (Figure 7). With specific site

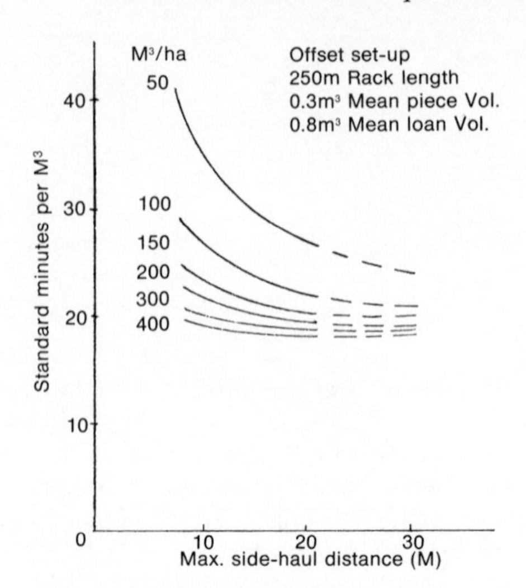


FIGURE 6. The effect of produce density

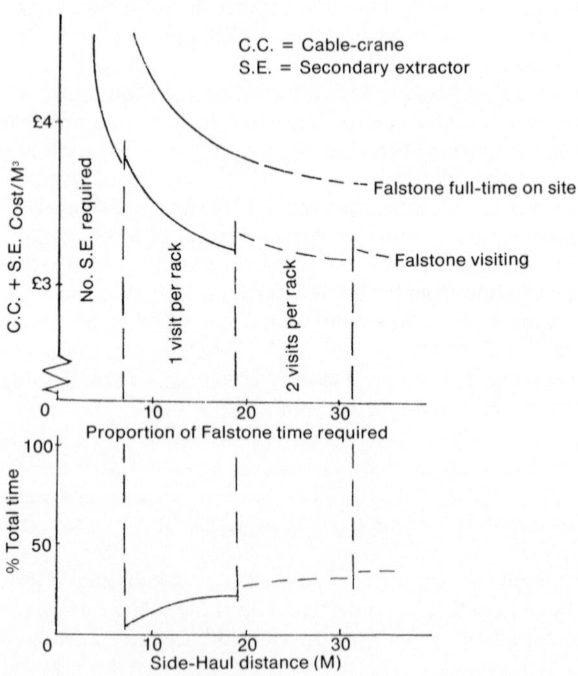


FIGURE 7. Cost on side-haul distance
Cable-crane and Falstone tractor—parallel extractors

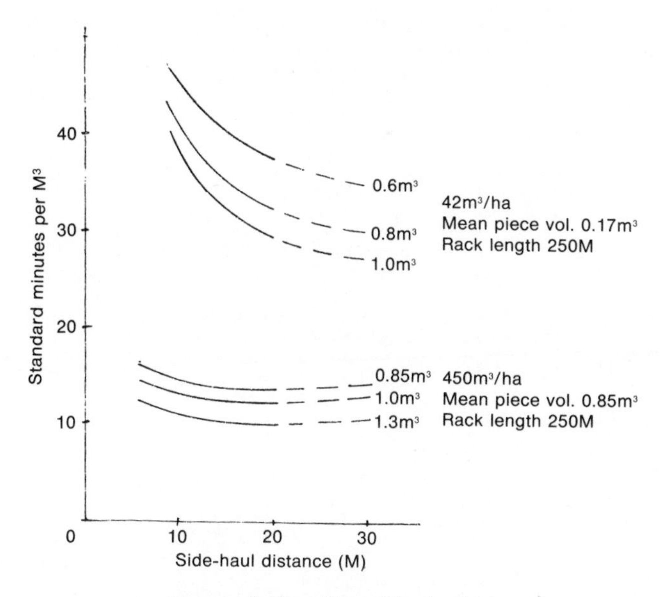


FIGURE 8. The effect of load volume

information including produce density, landing storage capacity and load sizes it is possible to estimate the number of visits required from a skidder, the mean duration of each visit and the proportion of the skidder time required to keep the cable-crane in operation.

3. Load volume (Figure 8). Tested over a wide range of crop conditions, it is evident that load sizes are critical if cable-crane extraction costs are to be minimised. There is no possibility that small loads can be extracted twice as fast as large loads.

4. Extraction Pattern (see Figure 4). Fan patterns from the tractor, often used to gather produce to a road end, are least efficient. Side-haul distances are high when the chokerman is furthest from the tractor and produce capacity on the landing is soon exceeded. The proportion of produce that must be moved by the secondary extractor is high. Fan patterns from the spar tree, where these are feasible, have a number of advantages. Longer side-haul distances are only encountered near the tractor where it is easier for the chokerman to pull the rope out. Produce is better distributed along the road, so the proportion of produce that must be moved by a secondary extractor is low. The number of visits required from the secondary machine is reduced to a minimum. Costs associated with the conventional parallel pattern of extraction tend to be higher than those for fan patterns from the spar tree, but lower than those for fan patterns from the tractor.

## DISCUSSION

Computer simulation can be useful for answering specific questions relating to harvesting methods or for routine estimation of machine interactions. This cable-crane model has been described only as an example of the type of question that can be answered by computer simulation and to illustrate the circumstances under which it might be considered worthwhile to use the approach.

SIMPLIFIED FLOW CHART—CABLE CRANE SIMULATION

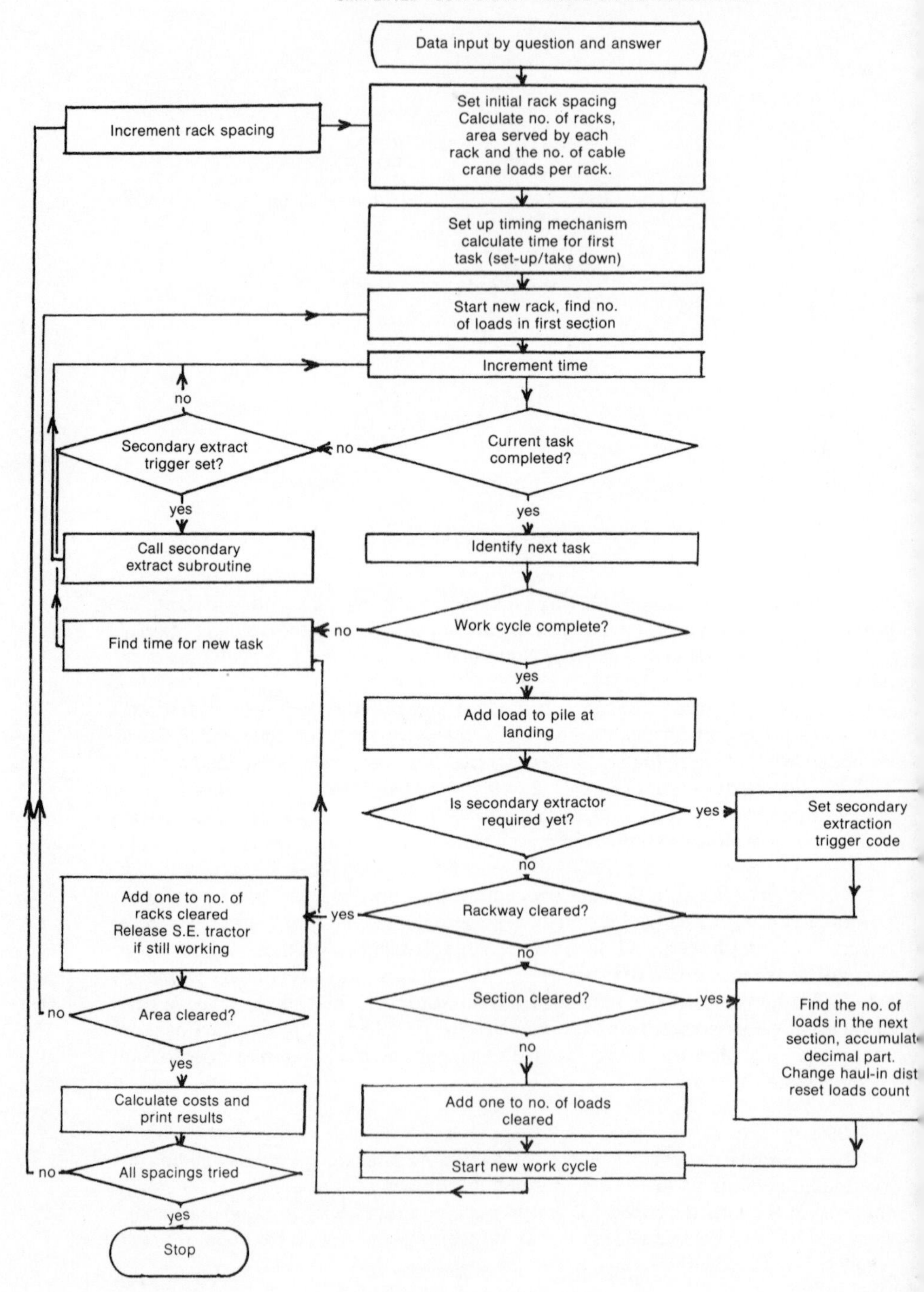

The objective was to create a tool for use by the Work Study specialist who wishes to try out a number of new ideas. The package has given general answers to a number of questions relating to whole-tree extraction, but needs detailed information on site and crop characteristics to produce useful answers. Future use is likely to be concentrated on extraction of shortwood produce from contour felling, and investigating the financial implications of creating additional roadside stacking bays at different intervals along the road.

Time study information must be available for each machine involved in any system, but provided the method of working is specified there is no requirement that the machines should ever have worked together in the real world. When theoretical optimal solutions are available, that is the time to try them out in practice.

ACKNOWLEDGEMENTS

The encouragement and help received from the Chief Work Study Officer and members of the Work Study Branch is gratefully acknowledged. In particular the author wishes to thank Mr B. Hicks for supplying data and posing problems for solving by simulation.

REFERENCES

Anon. (1978) Simulation techniques in Forest Operational Planning and Control.

Proceedings of Working Part S.04.01 IUFRO, Published by the Dept. of Forestry and Forest Products, Agricultural University, The Netherlands.

Forestry Commission Standard Time tables XXII/3, Whole Pole extraction by F.C. cable-crane (Skyline).

Hughes A.J.G. (1984) Harvesting Machine—Nomenclature. Forestry Commission Research Information Note 82/84/WS.

Jones, A.T. (1980) Skidder Based Secondary Extraction Systems. Forestry Commission Wales Work Study Team Internal Report No. 100.

# OPERATIONAL RESEARCH AS A PLANNING TOOL FOR HARVESTING TIMBER

ESKO MIKKONEN

*The Forest Work Study Section, Central Association of Finnish Forest Industries Metsäteho. Helsinki, Finland.*

SUMMARY

Theoretically, the goal programming approach seems to be the best routine for mathematically handling of problems related to wood procurement. The approach used must allow for considerable sensitivity analysis. If the effect of the variables expressed with different units is to be studied, interpretation of economic information given by the approach becomes a problem. The other drawback is that the approach does not allow determination of the hierarchy of the goals objectively as they depend on the subjective preferences of the decision makers.

From the practical point of view, standard linear programming is the best method if the desired objective can be formulated in economic terms. If there are several goals to be attained or satisfied the best method is goal programming. The more complicated and restricted the scope of the problem, the better the mathematical programming approach will be. This applies also in problems relating to harvesting. The solving of the problems would be impossible without efficient computers.

INTRODUCTION

The planning of wood purchasing is on the threshold of a new era in Finland. There are two reasons for this:

1) the change of wood purchasing organisations from line organisations to economically-independent decentralised divisions within companies;
2) the introduction of micro-computers in this field.

There has been rapid development of applications for operational research (OR) methods during recent years. This is partly due to the fast growth in the ability of computers to store and process data.

In the future, wood purchasing enterprises will require increasing capital inputs. Such operations are beginning to increasingly resemble high automated factories. The right allocation of scarce resources will therefore become more important. Furthermore, different options are restricted by law and agreements. Under these circumstances it is not feasible to neglect the opportunities offered by OR.

According to results of a questionnaire reported by Paakko (1983), OR methods were used by 31 companies in 109 applications during the 1970s. 68 percent of the applications were linear programming, 28 per cent simulation, and the remainder various heuristics. Methods were applied for medium term planning (45 percent), short term production planning (35 percent) and operation control (25 percent). The most important areas of application were production planning and control using mill models, and in economic planning, budgeting and financing. Within companies wood delivery and transport, sawmill operations, chemical forest industry applications, processing and energy models were most important. Other uses included research applications.

This paper will concentrate on the mathematical programming applications of OR because they seem to be the most applicable methods at present. In the following sections, the requirements for a good planning procedure will be defined. Based on this background, a few applications for wood harvesting and purchasing will be discussed.

## REQUIREMENTS FOR A GOOD PLANNING PROCEDURE

The ultimate goal for a good planning algorithm is to produce the best solution to the problem at hand. The optimal decision process consists of three steps:

1) One has to know, accurately and quantitatively, how the system is constructed and how the system variables interact.
2) There has to be a quantitative measure of effectiveness which is a function of the system variables. This can be profit, loss or some other economic measure.
3) The optimal value for the measure of effectiveness is found by the solution algorithm. Optimisation is decisive because it narrows down the possible choices to one—the best one.

Harvesting, transportation and processing of wood can be described as a linear economic model. The solution algorithm for this model has to be valid on two levels: it must be theoretically correct and the use of the algorithm must fit into the organisation which applies it.

Theoretically, the models and algorithms must:

include quantitative variables only;
produce feasible and optimal solutions;
be able to handle several decision variables simultaneously;
be able to take into account mutually contradictory goals;
include extensive sensitivity analysis;
give economic information on alternative solutions;
be able to handle variables measured with different measures.

From the practical point of view they have to:

be easy to formulate;
be applicable in real life conditions;
include the possibility of using time as a variable;
fit into the data processing hardware of the company;
be cheap to operate.

## DESCRIPTION OF THE WOOD PURCHASING AND HANDLING PROBLEM

The wood purchasing and handling problem can be formulated as a linear economic model, and include logging, handling, transportation and the handling of wood in the mill (Mikkonen, 1983). This approach allows one to look at the importance of several factors together; for instance, wood energy, rationalisation of different operations in the work chain, rational use of machine capacity and so on. Better investment planning is also possible through this approach.

The problem is usually restricted by the following factors:

Material constraints:
- work conditions or time;
- quantity or quality of raw material;
- machine or man hours available;
- productivity requirements.

Economic constraints:
- investment money;
- running costs;
- quality loss;
- transportation loss.

Constraints set by law and agreements.
Constraints set by the environment.
Production goals.

SOME MATHEMATICAL PROGRAMMING TECHNIQUES[1] (see also Table 1).

*Standard Linear Programming*

This method has the drawback that it permits only one objective function to be optimised. This is not the case in most real-life situations.

[1]Editors' note: Readers seeking further information about these techniques are referred to a general text such as Dykstra (1984).

*Parametric programming.* This method gives all the optimal solutions if done thoroughly. Unfortunately this becomes expensive and the interpretation of the results is a tedious task.

*Goal programming.* One feature of goal programming is that an otherwise infeasible problem can be reformulated so that it becomes feasible. This is done by introducing deviation variables and then maximising or minimising the sum of these deviations. One has to emphasise the fact that goal programming does not determine the hierarchy of the goals: this has to be done by the user.

The interpretation of the economic information of the goal programming solution is not clear if variables are measured in different units or the problem includes variables which cannot be expressed in economic terms. Proper interpretations of the results also require a knowledge of the hierarchy of the goals.

Consideration of properties of the algorithms (Table 1) would suggest that goal programming would be the most attractive option on theoretical grounds. Some practical limitations of the various methods are listed in Table 2 which indicate that goal programming or conventional linear programming would be the best alternatives.

Other possible techniques include Integer and Mixed Integer Programming, but for the purposes of this discussion, these are classed as special formulations of the goal or parametric programming approach and will not be considered further.

SOME APPLICATIONS

In one study (Mikkonen, 1978), a standard linear programming approach was used to determine the costs of the different alternatives in bunching of wood. Different methods were given subject weights allowing one to rank the alternatives. The linear programming approach worked well and it can be considered to be applicable for

TABLE 1. Theoretical properties of the solution algorithms

| Property | Std linear programming | Parametric programming | Goal programming |
|---|---|---|---|
| Quantitive | yes | yes | yes |
| Feasibility | yes | yes | yes[1] |
| Global optimality | yes | yes | yes[2] |
| Several goals | no | no | yes |
| Contradictory goals | no | no | yes |
| Integer variables | no | no | no |
| Economic interpretation | yes | yes | no[3] |
| Use of different measures | no | no | yes |
| Non-deterministic | no | yes[4] | no |

[1]In a mathematical sense the problem is always feasible.
[2]In a mathematical sense the solution is the global optimum.
[3]Possible in special cases.
[4]Produces ranges for all the solutions.

TABLE 2. Applicability of the algorithms for planning wood purchasing

| Requirement | Std linear | Parametric | Goal |
|---|---|---|---|
| Applicability to problem field | fair | good | good |
| Formulation | good | good | good |
| Time factor | good | good | good |
| Ease of use | good | fair | good |
| Interpretation | good | good | fair |
| Data processing | good | fair | good |
| Cheapness | good | bad | fair |

solving this kind of resource allocation problem. However, a more comprehensive sensitivity analysis should be performed.

In a study reported by Eskelinen *et al.* (1984), a linear programming model was developed for optimising the use of wood raw material when purchasing energy wood simultaneously with industrial roundwood. The model has been used by several Finnish forest companies, which were either starting to procure raw materials by whole tree harvesting and/or forest chipping methods, or were already doing it. The applicability of the model will increase, because of the improved capability of industry for accepting tree-lengths for pulping and fuel wood.

According to experience, the use of the model allows a more reliable determination of the economical volumes of raw material to be procured. It also allows better planning and choice of investments for both the mill and wood procurement organisation.

The third study dealt with the choice of an optimal saw-timber harvesting system (Mikkonen, 1980). The parametric programming approach gave the ranges for all the optimal values of the harvesting system under study. In conditions where the exact monetary value is not known, the approach gives the ranges for the decision variables.

In a fourth study (Mikkonen, 1983), a linear wood purchasing model of a large integrated forest industry company was built. The problem whether it is profitable to invest in a whole tree handling terminal at the mill was solved by using the goal programming approach. There were six different goal functions to be optimised during the course of the study.

## COMPARISON OF METHODS

1. Every method under consideration can be used as a planning procedure in the cost minimisation problem. When choosing the approach one has to consider whether there are several goals to be attained. If this is the case the most appropriate method is goal programming. If there is only one simple objective function, the best method is standard linear programming. The parametric approach is not an attractive option because of cost.

2. The parametric approach seems to be the best method for research into harvesting systems because it gives all the optimal solutions. However, modelling the complexity of real life systems can become very expensive. If there are several goals, goal programming is again the best method. There should be a possibility of using integer variables in the problem.

3. All the methods can be used for investment planning purposes. The parametric approach gives the range of the unit costs as a function of the size of the investment. If linearity does not hold, the non-linearity has to be linearised by using a proper method before further analysis.

## DISCUSSION

Operational research methods and mathematical programming are very capable and versatile problem solving tools. They can be successfully applied in harvesting related problems. There are many economic and technical constraints restricting the course of actions. A special feature for our problems is the biological forest environment within which they occur.

It is very difficult to master such complex systems without an efficient tool of analysis. Operational research is such a tool and should be applied more often.

## REFERENCES

Dykstra, D.P. (1984) *Mathematical Programming for Natural Resource Management.* McGraw-Hill.

Eskelinen, A.; Häggblom, R. and Peltonen, J. (1984) A model for optimising the use of wood raw material when procuring energy wood in conjunction with industrial roundwood. *Metsäteho Report No. 389.* Helsinki.

Mikkonen, E. (1978) Development alternatives in the bunching of timber. *Metsäteho Report No. 352.* Helsinki.

Mikkonen, E. (1980) The choice of the optimal sawtimber system. Unpubl. *Master of Forestry Thesis.* Oregon State University, USA.

Mikkonen, E. (1983) The usefulness of some techniques of mathematical programming as a tool for the choice of timber harvesting system. *Acta Forestalia Fennica No. 183.* Helsinki.

Paakko, J. (1983) OR methods and applications in the Finnish forest industry. The 1st NOAS-seminar proceedings. *OR in Forest Industry.* Mimeograph. Helsinki.

# INTRODUCING COMPUTER SYSTEMS THAT IMPROVE DECISION MAKING

MARK R. LEMBERSKY

*Group Systems and Finance Director*
*Weyerhaeuser Company, Tacoma WA 98477, USA*

SUMMARY

One of the major opportunity areas for applying computer systems to forestry is using them to improve operational decision making. This paper is about easier and faster introduction of such systems.

The continuous raw material flow from forest land to and through the mills creates many complex operational situations. These require strategic decision making by managers and moment-to-moment tactical decision making by field, yard, and mill workers. Decisions involve harvest scheduling, raw material transportation and distribution, log allocation among alternative end uses, and many other areas that will immediately come to the reader's mind.

Many computer-based quantitative methods and analytical procedures have been and will be developed with the goal of helping improve managerial and/or field worker decision making. The most difficult and time-consuming part of successfully implementing such methods and procedures generally is *not* technically analysing the particular situation under investigation, nor is it creating the computer software to carry out the necessary calculations. Rather, the most challenging task usually is earning credibility and achieving acceptance by the managers and workers who are the potential users and beneficiaries. Many of these individuals have not had prior exposure to computer-based decision making tools and perhaps are sceptical of their worth. This paper describes an approach, based on 'Decision Simulators,' that has been quite successful over the past eight years in addressing these questions for both senior managers and field workers at Weyerhaeuser Company—and that I feel has applicability well beyond our company. The approach makes it easy for the manager or worker to understand and explore the quantitative method of analytical procedure *without* needing to develop any computer expertise or learn new technical methodology and jargon. A Decision Simulator involves an easy to use, highly interactive, *visually oriented* computer representation of the operational situation under consideration. A Decision Simulator permits and encourages its user to explore decision making alternatives and to discover on his or her own the best solution to the problem at hand.

What follows is an edited and somewhat modified version of 'Decision Simulators Speed Implementation and Improve Operations,' by M.R. Lembersky and U.H. Chi, previously published in *Interfaces* 14:4pp 1-15 (July-August 1984). The reader interested in more general discussion of Decision Simulators is referred to the original paper. This version is printed by permission of the authors and of the journal and is 

## INTRODUCTION

The challenge in implementing computer-based tools to aid decision making is often not in developing a technically sound method, but in generating credibility in the minds of potential users. Here we refer not to the credibility of the solution developers, although this is obviously a prerequisite. Instead we refer to the decision maker's confidence in the specific decisions or actions suggested by the method. Often implementation is accomplished by simplifying the method to the extent necessary to gain user understanding and credibility. This approach is usually justified and appropriate. However, there are many situations where either: (1) the full power of the method is needed to adequately address the decision problem, or; (2) the amount of profit or cost savings foregone by the simplified approach is large.

When such situations involve operating decisions, presenting the underlying method in a Decision Simulator (DS) can be effective. A DS is not a new analytic procedure, but an approach to implementation that increases the likelihood that the underlying tool for decision making will be used. Like video games and flight simulators, a DS provides an interactive, visual (instead of numerical) simulation of the actual decision-making process, including the consequences of the decisions made. Using a DS, the decision maker can observe, experiment with, test, and even try to outperform the computer generated solutions in an environment with which he is already familiar. During the eight years we have utilised DSs, we have seen users quickly come to terms with the decision tool's performance and its solutions readily gain credibility.

Our initial DS was designed in 1976 and put into practice during 1977; this work was held proprietary until a patent covering several of its key elements was issued (Chi and Lembersky 1982). We describe this DS in some detail, because we believe the general idea is best conveyed by example. Further, the application continues to produce several million dollars annually of increased company profits and is of interest in its own right. We also identify key DS features for use by readers in other situations.

## TIMBER PROCESSING DECISIONS

In commercial forest operations mature trees are harvested, delimbed, and often topped. The resulting 'stems'—often 60 to 100 feet long—are crosscut into logs of various lengths. These logs are allocated among different mills, each mill manufacturing a different end product (for example, lumber, plywood, or paper). For each stem there may be hundreds of reasonable combinations of log lengths that could be cut and allocated. Each combination results in a different set of end products and, therefore, a different set of revenues. Consequently, these crosscutting and allocation decisions affect the profits realised from the tree profoundly. Because each tree is physically different from every other one, it is not possible to find a 'right answer' to apply to every stem. Rather, these decisions are made stem-by-stem hundreds of times each day by geographically dispersed operators. These decision makers must simultaneously consider a stem's geometric profile (including length, diameter, taper, and curvature), quality variations along the stem (including knots and rot), and varying economic utilities of potential logs of different geometries and qualities in different mills. This is a formidable task for the operators, especially under the circumstances where these decisions are usually made: in the brush of the forest floor or in the operator's cab of a large mechanical facility where six or more stems may require cutting and allocation per minute. A consistent improvement in this decision process, even though relatively small, can produce a major increase in the company's profitability.

For any given stem it is possible to complete a cutting and allocation solution via a procedure called 'dynamic programming' that maximises return for that stem.

Indeed various versions of such a dynamic programming method have been discussed in the literature, starting with Pnevmaticos and Mann (1972). However, the impact on day-to-day operations was minimal. Improved decision making was not transferred to field operators. Solutions resulted from an unfamiliar, somewhat intimidating, 'black box' procedure.

### VISION SYSTEM

We tried to change this situation in 1976 using the decision simulator approach (Lembersky 1977; Chi 1977). We sought to achieve user credibility in the dynamic programming solutions by allowing operators to deal with them on their own terms; that is, in a visual context that is familiar and non-threatenting. We sought a system that built on the users' considerable experience and increased their decision-making skills by giving them feedback on the economic impact of their cutting and allocation decisions compared with the dynamic programming solutions.

The system we developed simulates all salient features of the actual decision-making situation, while providing a private environment for 'non-destructive' learning by doing. The operator sees on a video display a realistic representation of each system. The DS permits the operator to inspect the stem (for example, roll and rotate it) and then cut and allocate it. The operator sees the logs and the profit achieved by his decisions with each stem and he also sees the stem cut and allocated by the programming procedure so as to maximise profit. The operator can recut the same stem repeatedly to explore alternate decisions.

These capabilities exist as a family of three versions, differentiated primarily by level of realism and corresponding hardware cost. The essential elements of the system are contained in all three versions; indeed, a design objective of each successive version was to reduce the hardware costs while retaining these essential capabilities. The original version is called VISION (Video Interactive Stem Inspection and Optimisation). Its 'descendants' are Mini-VISION and Micro-VISION, the prefix referring to the class of computer hardware utilised.

VISION utilises high performance, 3-D interactive graphics hardware directly linked to a powerful host computer (such as a DEC VAX 11/780 or a Harris S500). Hardware costs were approximately $300,000 in 1976. Mini-VISION is a stand-alone table-top system utilising a Tektronix 4052 computer, or equivalent hardware, at a cost under $30,000. Micro-VISION, including a self-contained tutorial, is currently designed for the IBM Personal Computer, costing under $3,000; a two times order of magnitude cost reduction from the original.

### TYPICAL VISION USER SESSION

A visually oriented, dynamic, interactive system cannot be adequately described with words on a page and illustrations. Nonetheless, we shall try to convey the design and operation of the VISION family by describing a typical VISION user session:

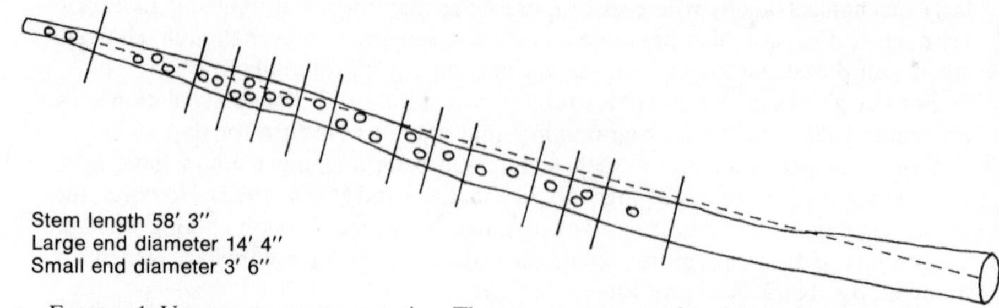


FIGURE 1. VISION stem representation. The crosses correspond to changes in stem quality and the small circles indicate knots. (The small circles are here used to enhance photo reproduction; knot indicators appear differently to the VISION user.)

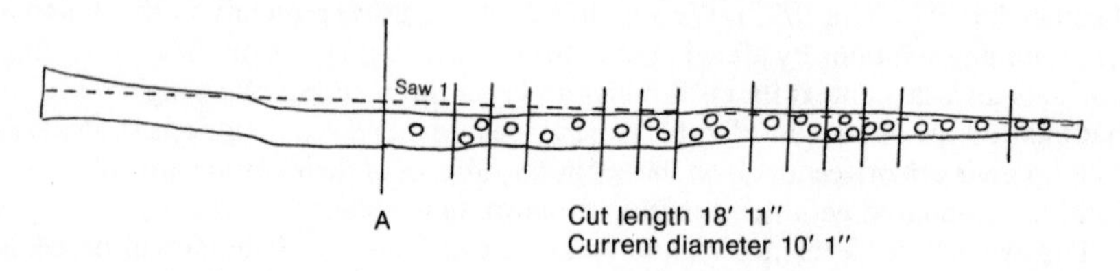


FIGURE 2. Simulated stem cross-cutting. The saw (A) has moved from its position and is shown cutting through the stem at user-selected length.

1. Start-up: The user selects a set of stems and the set of manufacturing facilities that will receive the cut logs. We gathered and stored stem geometry and quality information for large, representative samples of stems from our operating regions. In addition, VISION allows the user to create stems by drawing their profile on a data tablet. We also stored the value of logs of different lengths, diameters, curvatures, and quality characteristics for each of our manufacturing facilities.

2. Stem inspection: A stem is displayed (Figure 1). The stem can be rotated, rolled and generally reoriented in real time using a joystick. The user can see a stem from several perspectives and, for example, discover hidden curvature. The display can be set so each stem initially appears on the screen in a specified orientation, reflecting the user's view in actual practice. The user inspects the stem until ready to make decisions.

3. Crosscutting and allocating: The user cuts the stem and allocates the logs to the mills. The display for this step varies with the type of operation simulated. For example, if the user normally works in the woods, he or she moves a 'saw' to the desired cut length along the stem and then applies the saw (Figure 2). For a multiple saw mechanical facility, the user punches a button for each saw selected, as is done in an operator's cab, and allocates the logs by pressing appropriate buttons.

4. Feedback: Based on the logs the user produces and their economic value to the mills where they are sent, the system calculates the value of each log and displays those values along with other information about the logs. It also displays the sum of the log values. Thus, VISION shows the economic consequences of the user's decisions (Figure 3).

5. Comparison: The system next uses dynamic programming and log value data to

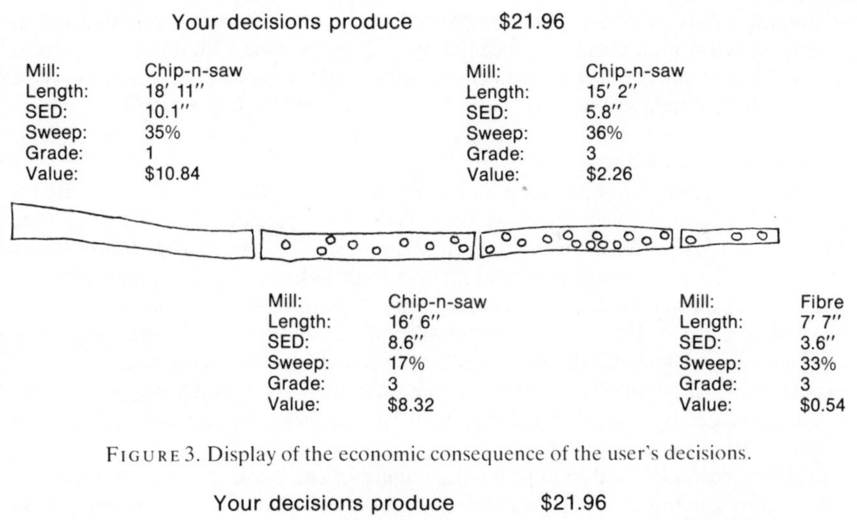


FIGURE 3. Display of the economic consequence of the user's decisions.

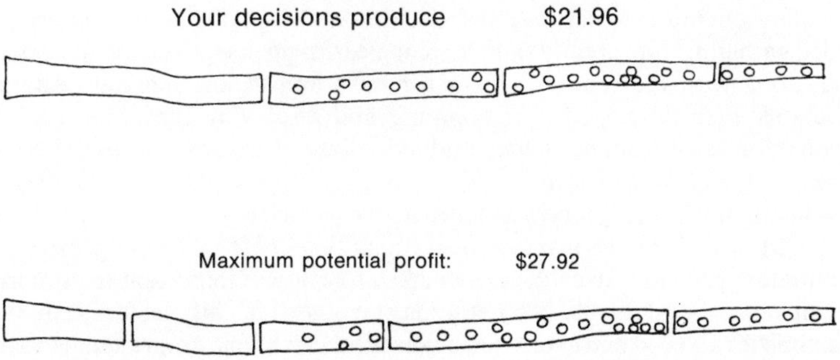

Your decisions produce $21.96

Maximum potential profit: $27.92

FIGURE 4. Maximum profit decisions displayed for comparison with the user's decisions.

compute and then display a method of cutting and allocating the stem which maximises the stem's profit contribution (Figure 4). The computations begin while the user is inspecting the stem. The user learns by comparing his decision with the maximum profit-producing one. Often, after observing where particular decisions lose or gain value, a user cuts and allocates the stem again.

6. Evaluation: After finishing with that stem, the user proceeds to the next and repeats the inspection, decision, and comparison steps. This continues until the user either ends the session or requests a cumulative "score card". This score card compares total value obtained by the user with the maximum possible for stems handled during the session. After noting the score card, the user can process additional stems.

A Mini-VISION or Micro-VISION session follows the same sequence. Initial stem file and mill selection occurs by inserting an appropriate cassette (Mini-VISION) or diskette (Micro-VISION). The current hardware configuration of both the Tektronic and the IBM PC dictate that the alphanumeric keyboard be present, but only the function and numeric pad portions are used. The main difference from VISION are that stem inspection does not include three-dimensional rotation or roll (stem orientation is fixed in the side view that exposes maximum curvature) and maximum profit decisions are precomputed and stored along with the stem data. The latter requirement reflects 1983 capabilities of small computers—it seems likely that within a year or two complex real-time calculations will be possible with micro-computers. Three-dimensional roll and rotation capability is a bit further off, but will certainly be possible before long in low-cost hardware, as indicated by Clark (1982).

### TYPICAL VISION APPLICATION

The VISION family has been used in Weyerhaeuser Company since 1977. Often a group of operators will build decision skills by spending time with VISION. Management uses VISION to evaluate the economics of alternative stem processing strategies. The most interesting uses involve both individual training and strategy changes. A typical application of the latter type involves very high-value mature West Coast Douglas Fir. These trees are generally cut and allocated in the woods because the logs go to geographically separated locations—to a conversion facility or to an export dock, for example. Operators make their stem-by-stem decision using their judgement aided by general strategies, called 'woods bucking instructions', printed on a pocket-sized card.

VISION was both used to improve the quality of the woods bucking instructions and to allow operators to improve their decision-making skills. A development phase included gathering appropriate sample stem measurements for each major operating region along with the economic data for mills and export markets. Alternative instructions were developed by computing and displaying the profit-maximising decisions for each sample stem, studying these decisions for patterns which reappeared for classes of stems, and attempting to generalise these patterns into woods instructions. This process included active participation by a respected veteran field expert who brought his vast operational experience and judgement to the generalisation process. Given the expert's key role, it was important to establish the credibility of the dynamic programming method in his mind. VISION's DS characteristics allowed him to test and challenge the dynamic programming solutions and convinced him of the value of the procedure's results. The visual and interactive nature of VISION enabled him to construct alternative bucking instructions easily and compare alternative cuttings patterns. The expert commented that it would have been very difficult and taken significantly longer to develop new instructions by looking at printed numerical output. Once a number of these candidate instructions were developed, each was applied to all the sample stems and the resulting aggregate profits compared. The best performing set of instructions was selected.

After the new woods bucking instructions were picked, VISION was used to train the woods foremen in their interpretation and use. In this phase, the foremen were first asked to cut and allocate a sample of stems using their old instructions. Then, they were given the new instructions and asked to cut the same set of stems using these instructions. The foremen were also encouraged to experiment with VISION. The foremen readily embraced the new woods instructions and understood the utility of the dynamic programming method. As a result, their operating regions implemented the new instructions almost immediately.

The operational results were carefully tracked at a number of field locations. The measured benefit from the new instructions and related VISION training was $7 million per year. Other VISION applications have impact of similar magnitude.

## DECISION SIMULATORS

We now turn to a general characterisation of DS systems. VISION and the other DSs we have worked with each have the interrelated features listed below. The features indicate 'what one observes', 'what one can do', and 'how one can do it'.

1. A DS provides a believable representation of the actual decision-making environment. This does not require detailed replication, but just enough realism to be credible to the ultimate user. It is also helpful to present only that portion of reality which is appropriate to the problem at hand. For exmple, VISION uses a 'wire-frame' representation of each stem because it communicates what the dynamic programming procedure 'sees'; that is, diameters and offsets from a centerline at appropriate intervals. A more detailed presentation of the stem, while making for a more complete picture, would obscure the fact that the optimisation method did not utilise the additional detail in its decision computations. Balancing the objectives of realism for the user and accurate portrayal of the underlying decision tool is an important challenge in designing a DS. The visual representation should include real-time dynamics and, where appropriate, animation. This aids in creating a believable decision-making environment.

2. DS is highly interactive and provides immediate feedback on the effect of decisions. These characteristics are very important to both user acceptance and user learning of better decision making. We follow the guideline that DS actions which correspond to 'real world' actions should exhibit response times comparable to their real world counterparts. A DS also emphasises 'seeing' the impact of decisions, rather than displaying data about the decisions (as in tabular or bar charts).

3. A DS is easy to use, without special training. The goal is systems that the typical user can successfully operate with a minimum of instruction and with little 'user support'. Almost every good video game demonstrates that this is achievable in quite complex situations. VISION users were proficient after five minutes of verbal instruction and rarely required further user support. DS users should not need to learn new specialised skills or a special vocabulary (such as command languages and their associated syntax) nor deal with devices (such as keyboards) that are not already part of their operational experience.

## ISSUES AND OPPORTUNITIES IN DS DESIGN—A WORD TO THE DS DESIGNER

We feel the central issue confronting the practitioner creating a DS is selecting appropriate visual and interaction abstractions. We are currently investigating various abstraction mechanisms for implementing DSs in alternative application domains. Our view is that while each application area may require some visual 'tailoring' we expect to find common visual and interactive mechanisms that are useful for related problems (for example, flow simulations for different situations). Such common mechanisms would simplify the design of DSs.

Other potential issues, such as hardware cost and dependence, difficulties in graphics programming, and lack of physical portability are at worst transient technical phenomena which become less important as computing resources continue to become more powerful and less expensive. This is illustrated by the rapid evolution of VISION from large, specialised $300,000 hardware to an IBM PC.

Some tools are evolving that potentially can aid in DS construction. For example, Bergman and Kaufman (1976) and Kaufman (1978) discuss computer language for programming DS-like real-time graphic systems. Withers and Hurrion (1982) describe an approach for interactively constructing visual simulations. Also see Borning (1981) and LaPalme and Cormier (1983).

Finally, technological advances will allow DSs of greater capabilities in terms of realism, real-time interaction, and data storage. Some of the possibilities (in a video game context) are described by Perry (1983).

### FINAL COMMENT

Prior to the availability of low cost, high performance computer hardware exemplified by the personal computers on the market today, the DS approach was viable only for organisations with significant resources to spend on such an endeavour. Now the approach is feasible for a much larger group of forest industry problems. The benefits of following this approach can be considerable.

### ACKNOWLEDGEMENTS

Doug Hay prepared in 1975 the dynamic programming procedure later used in VISION. Billy Chou and Jay Gischer provided excellent software support. Bill Grunow played a key role in the 'woods bucking' application.

### REFERENCES

Bergman, S. and Kaufman, A. (1976) BGRAF2: A real-time graphic langauage with modular objects and implicit dynamics. *Computer Graphics* **10**(2)133-138.

Borning, A. (1981) The programming language aspects of ThingLab, a constraint-oriented simulation laboratory. *ACM Transactions on Programming Languages and Systems* 3(4)353-387.

Chi, U.H. (1977) The VISION system. *Weyerhaeuser Internal Document 044-6101.*

Chi, U.H. and Lembersky, M.R. (1982) Simulated interactive dividing and allocating process. *US Patent 4364732.*

Clark, J.H. (1982) The geometry engine: A VLSI geometry system for graphics. *Computer Graphics* 16(3)127-133.

Kaufman, A.A. (1978) System design and implementation of BGRAF2. *Computer Graphics* 12(3)87-92.

LaPalme, G. and Cormier, M. (1983) Design considerations for an interactive route editor. *Publication No 319,* Centre de Recherche sur le Transports, University of Montreal, Quebec, Canada.

Lembersky, M.R. (1977) Interactive video display package. *Weyerhaeuser Internal Document 044-2656.*

Perry, T.S. (1983) Video games: the next wave. *IEEE Spectrum* **20**(12)52-59.

Pnevmaticos, S.M. and Mann, S.H. (1972) Dynamic programming in tree bucking. *Forest Products Journal* 22(2)26-30.

Withers, S.J. and Hurrion, R.D. (1982) The interactive development of visual simulation models. *Journal of the Operational Research Society* 33(11)973-975.

# PART V

# COMPUTERS IN THE WOOD PROCESSING INDUSTRIES

# THE MARKETING OF ROUNDWOOD

P.J. FOTHERGILL

*Forestry Consultant, 14 Queens Square, Bath BA1 2HN*

SUMMARY

The advent of the British Telecom 'Prestel' information service coincided with a period when it was difficult to find markets for timber. The author has developed an Information Handling System for the British forest industry which is based upon Prestel and is designed to make marketing easier.

The system provides a countrywide service using straightforward, easily accessible information. Among subjects covered by the database are details of timber for sale, timber required and equipment for hire. The costs of using the system are low, with only simple hardware being required.

## BACKGROUND AND THE PRESTEL SYSTEM

### *The original idea*

In 1982, Information Technology Year, the UK forestry industry found itself struggling with the problem of the marketing of softwood thinnings and continued mounting costs. The exchange rates at that time were such that the pound was expensive in relation to other currencies. In turn this had led to the closing of some major pulpwood mills in Britain, with a consequent countrywide difficulty over the disposal of small diameter softwood thinnings.

In the southwest of England, where I happen to work, the softwood thinning market continues to be a headache to woodland owners and managers, and the Timber Growers United Kingdom (TGUK) have been doing all they can to improve the working of the market and find new outlets.

It occurred to me that the main problem so often is the lack of *easily available information* for both sides of the market. This applied not only to the softwood thinnings market, but also to the general sale and purchase of timber.

### *The Prestel system*

Information technology has developed with great speed, and many of the advances are British inventions. One of these is Prestel[1]. Prestel has been developed by British Telecom, and provides an information service via the telephone shown on a television screen.

Stated simply the Prestel service involves the display on television screens or visual display units of written or graphic information, sent via the telephone network from computers set up to carry databases of information for a large variety of purposes.

The Prestel system provides facilities for individual people, companies or organisations to supply information to form a database, placed on the Prestel

[1]Prestel is the trademark of the British Telecom Prestel System.

computers, to which users can have access through their own receiving units. These receiving units can be either purpose-made Prestel receivers, or a television set with a suitable decoder which can translate the information coming over the telephone line so that it can be fed directly into the television receiver.

Anyone in the country who has a telephone can use the Prestel system, either by having a Prestel receiver or an adaptor to their television set. A simple 0 to 9 keyboard allows information to be 'dialed up' and displayed on the screen. An alphanumeric keyboard will, in addition, allow messages to be sent and information exchanged.

The information on the Prestel service is in the form of individual pages or 'frames', each carrying 960 characters of displayed information. In addition, each frame holds another 64 characters which are used by the Prestel service to control information, such as the location of 11 other frames or pages to which the viewer may be routed directly by keying a single character (0 to 9 or #) on his keypad. Users can therefore be offered a choice of frames and can be routed to different items of information by using this system of selection.

The Prestel system is run through British Telecom who rent pages on the computer to major organisations which are contracted to be what are known as Information Providers (IPs). These IPs act as 'wholesalers' hiring a large number of pages and organising the distribution and allocations of these to smaller organisations who need a number of pages.

These smaller organisations are known as Sub-information Providers (Sub IPs) and they in turn rent their pages through an IP.. The Sub IPs, who set up the individual databases, are permitted to amend and update the pages allocated to them and the pages are secured to them and cannot be tampered with by other persons.

There are already some 329,000 individual pages of information available, any of which can be called up by the user. Most of the pages on the Prestel service are available for any user to inspect on the screen. It is possible, however, to have what is known as a Closed Use Group (CUG). Pages coming under the control of a CUG can only be viewed by persons authorised by the CUG to see them. This is ensured by a system of safety codes. Information can therefore be put on a database for restricted retrieval by only people authorised to retrieve it.

The Prestel service was initially slow to grow, after its inception 6 years ago, but it began to be used for the United Kingdom Financial Market and for the holiday and travel industry, and is now being used considerably by British lawyers for legal information.

I thought that the British forestry industry could also make use of Prestel. What could Prestel do for the forestry industry? First, owners and managers could use the system to advertise the timber they had available, and second, timber merchants could use it to make known the timber they were searching for. There would be no need to disclose price levels, only species, diameter and quantity, location and basis of purchase.

## THE FORESTEL AND TIMBERTEL SERVICE

### *The new service*

In 1983, I set up a trial system on Prestel under the two headings:

FORESTEL: for forestry information

TIMBERTEL: for the timber market

TIMBERTEL operated as a subsection of FORESTEL. I obtained some 80 pages on

Prestel to enable me to provide a fairly comprehensive range of information through an IP.

Once I had worked out the details and discovered the benefits and problems, I passed the system to the Timber Growers England and Wales (now amalgamated with Timber Growers Scotland as TGUK).

TGUK now operates the FORESTEL and TIMBERTEL Service, in close cooperation with the British Timber Merchants' Association, and I have been doing their editing, and dealing with the placement of all information on the screen.

*Benefits of the system*

The FORESTEL and TIMBERTEL Services on Prestel have the particular benefits that:

a) Any information placed on the screen is viewable throughout the United Kingdom.
b) The information can be placed on screen immediately. (I usually deal with it in 24 hours.)
c) Alteration or updating of information on screen is easy and immediate.
d) The provision of the system and its use are very inexpensive.

*Example of the use of the FORESTEL system*

Let us suppose that a woodland owner wishes to offer a parcel of thinnings for sale. The owner will normally select the trees which are to be sold, mark them in some way, and assess their volume. To contact markets, he would normally take some or any of the following actions:

a) Advertise in national forestry journals.
b) Advertise in local newspapers.
c) Advertise through a specialist firm dealing with timber sales.
d) Contact an organisation such as TGUK for advice on possible markets.
e) Write to or telephone a number of timber merchants thought likely to be interested in the particular species and size of timber which are for sale.

Possible purchasers would then write to or telephone for more detail, and to arrange to inspect the timber and discuss prices and basis of sale.

Now that FORESTEL is avialable, there is a speedy alternative, since, given an appropriate viewer or adaptor, he can either:

f) View the FORESTEL Service (TIMBERTEL section) and look at the pages of Timber Required. If there are timber merchants advertising for the type of timber which he has on offer, then he can contact the merchant direct, or
g) He can place an advertisement on the Timber for Sale section of the TIMBERTEL or Timber Market part of the FORESTEL Service. This is done by contacting either the editor of FORESTEL, or TGUK, or the British Timber Merchants' Association.

The owner will need to give the following information:

a) The name of the vendor (e.g. the name of the estate).
b) The location of the timber (nearest large town).
c) STD code and telephone number to contact owner.
d) Timber species.
e) Timber volume in cubic metres.

f) Timber minimum top diameter (cm over bark) or diameter at breast height (cm over bark).
g) Basis of sale, e.g. standing, felled, at roadside or at yard.

The information will then be placed on the FORESTEL database, and be viewable throughout the United Kingdom, normally within 24 hours.

Timber merchants can view the Timber for Sale offer on screen either directly, or through their Timber Merchants' Association (who have viewing equipment at their headquarters), and contact the owner by telephone to discuss purchase details and prices.

In a reverse way, a timber merchant can use the system to advertise Timber Required, and make initial contact by telephone with an owner selling timber of the type required.

*How it works*

a) Information coming in is passed to the editor.
b) The editor uses his Prestel terminal to contact the central Prestel computer.
c) The editor keys in the text of the new information.
d) The Prestel central computer passes the same details to all the regional computers.
e) Anyone, anywhere in the UK, with a Prestel terminal can dial up the information pages.
f) Advertisements on screen always carry a telephone contact number, and contact can then be made direct by an enquirer to the advertiser.

*Dialling in to the Prestel Service*

This is done by dialling via your terminal to the Prestel computer nearest to you.

Once your terminal is in touch with the Prestel computer, if you know the page number, you can obtain it at once; if not you follow through a general index until you reach the information you require.

*Present costs of advertising on the FORESTEL Service*

The system provides information at low cost. The costs (1984) are:

| Period | Per page | Per 3 lines | Per each additional line | (38 letters/line) |
|---|---|---|---|---|
| Monthly | £40 | £10 | £4 | (18 lines/page) |
| Quarterly | £60 | £15 | £6 | (18 lines/page) |
| Yearly | £120 | £30 | £12 | (18 lines/page) |

Any change during the period is charged at 20p per line altered.

*The hardware needed*

If you wish to have a terminal in your office or home, then the following table gives an idea of approximate costs. Lower prices are for black-and-white vision or numeric keyboards, whereas higher prices are for colour and alphanumeric keyboards.

| | *Purchase* (£) | *Hire* (£ per year) |
|---|---|---|
| Terminal complete | 99-550 | 95-226 |
| Adaptor for TV | 129-225 | 66-96 |

*Viewing costs*

To view requires a local telephone call costing either:

a) 4p per minute from 8.00am to 6.00pm weekdays and 8.00am to 1.00pm Saturdays, or

b) nothing at other times.

Sensible choice of times and care in use can keep costs down (e.g. dialling in to a page direct instead of via a series of indices). The more this service is used the less the costs will become.

*What is 'on screen' at present*

A wide variety of forestry information is available on FORESTEL, including:

Timber for Sale
Timber Required
Forest Nurseries
Forestry News
Forestry Commission Auctions
Forestry Consultants
Forestry Bodies

More useful information will soon be put on screen to help the industry, including:

A map of Forestry Commission areas
An index of their addresses
Tables of tree spacing for quick reference

MAILBOX AND RESPONSE PAGES

Mailbox is a system set up by Prestel for use by any persons who have a Prestel receiver. It enables them to send messages for collection by the recipient to whom the message is addressed the next time that recipient switches on his Prestel receiver. Safety codes incorporated in the system safeguard the messages for the correct recipient only. The messages are then stored until they are seen by the recipient, who then has the option to put the message in storage or erase it. Up to 14 messages can be put in store for any one recipient at one time at present. An increase in the number is planned by Prestel.

The cost of using the Mailbox service involves the cost of the local telephone call made to Prestel (the charge for this varying according to the time of day as with all ordinary telephone calls). There is no special charge made for using the service, over and above the normal rate charged for the use of Prestel, and the telephone call to connect with the Prestel computer.

For two-way communication, and to encourage simple and rapid response, Prestel provides a system known as 'Response Pages.' A response page is one in which a form is shown on the screen, which can be filled in by the viewer and returned immediately to the advertiser. These are more expensive to hire than ordinary pages, but have special value in making available such things as order forms, returns or brief reports, without the need for them to be sent out first by post. At present response pages are not in use with the FORESTEL Service, but will be brought into use when the throughput of advertisements increases, and when there are more Prestel terminals in use.

## THE FUTURE

### *Increased use of the existing FORESTEL Service*

The system now has some 60 users, and the use has grown as the industry has taken up the idea. It is difficult at present to tell exactly how many offers have been made and accepted by users of the system, as feedback is sporadic and the present number of users too small to allow any logical deductions on the rate of expansion of the use of the FORESTEL Service. However, if the present rate of use of the Service and its growth are to be taken as a guide, then an increase of about 10 additional advertisers per month would seem likely.

### *Expansion of the Service offered*

An important expansion of the facilities offered at present will be the linking in of the FORESTEL System into one of the databases which exist to assist with freight handling. This would enable users of the FORESTEL Service to arrange for the transport of timber. Such a link is at present under discussion.

# COMPUTER AIDED SAWMILLING—RESEARCH FOR BRITISH SAWMILLS

K.W. MAUN

*Princes Risborough Laboratory (BRE), Princes Risborough, Aylesbury, Bucks*

SUMMARY

Optimising programs, which approximate sawlogs to truncated cones, have been designed and used to help British sawmills to improve the yield from a batch of logs. However, truncated cone models are not ideal for logs which can have irregular cross-sections and are often bent. Therefore, PRL has developed and tested a unique computer optimisation model which allows the actual shape of a log to be simulated accurately.

Models have only been used to date in a 'background' mode to help in the design of sawmills and to produce data on how a batch of logs should be cut up to satisfy sawn timber markets. However, further improvements in the yield from sawlogs can be gained by using an 'on-line' computer system which controls the conversion of individual sawlogs. A system has been developed for this purpose and can be used in both existing and new sawmills of varying designs.

In an extended trial in an industrial sawmill, it improved the control over the products cut and increased the total yield of sawn timber by five percent.

INTRODUCTION

Most UK sawmills produce timber for a specific purpose—for example, building and construction or mining or pallet manufacture. Their primary products, which command the highest prices, are battens or boards of a number of preferred cross-sectional sizes and lengths, designed to meet these markets. Secondary products, commonly random width and length boards, are also produced but are usually less valuable and are sold for different purposes.

Traditionally, sawmills have relied upon the experience and skill of the operators in order to convert sawlogs into the products which they require. However, the decisions taken are not always optimal, mainly because the operator cannot logically take account of all the important factors which affect the efficient conversion of a log. His task would be easier and the logs converted to a consistently high yield if a computer decision making process were used.

In recent years there has been considerable interest in and research into the extended role that computers can play in the sawmilling industry. Particular attention has been paid to improving the yield and control of sizes of sawn timber cut from logs by using computer aids. The Princes Risborough Laboratory (PRL) has been very active in this field and has developed a range of simulation programs for main-frame and micro-computers. These have been used at the Laboratory in a background mode (i.e. not 'on-line' at a sawmill) to develop cutting patterns and to investigate conversion procedures for sawmills. Also, they have been used in research which has

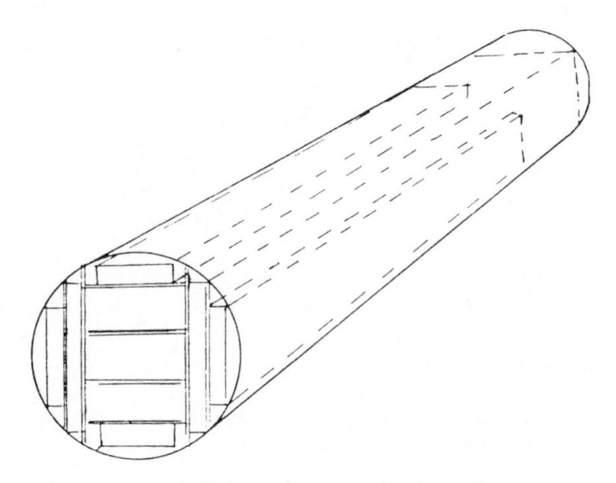

FIGURE 1. Stylised cutting pattern for truncated cone.

led to the development of an 'on-line' system for the controlled conversion of sawlogs known as LOCAS (The Laser Optimiser and Cant Alignment System). LOCAS simulates alternative ways of converting a log in order to satisfy particular markets and after analysing the resulting yields for each conversion pattern, selects the optimum. This is achieved by a mathematical model which represents the complete log shape, the cutting procedures of a sawmill and the market requirement for various sizes of sawn timber. The programs optimise on the basis of the volume of preferred sizes and can be used either to satisfy actual orders or to satisfy stock level requirements of those sizes. An optimisation on the basis of the value of the sawn out-turn can also be carried out, but since it is not at present possible to predict the quality of the sawn timber, the value of particular pieces would be based upon their dimension.

## BACKGROUND OPTIMISATION PROGRAMS

### *General*

The use of sawlog conversion optimisation programs is becoming increasingly more important as capital investment in sawmills increases and as the cost of logs increases. They can be used in a background mode to generate stylised cutting patterns, see Figure 1 or to simulate the sawn out-turn from a batch of sawlogs. The information gained can be used to aid the design or modification of a sawmill and to help towards maximising the yield of timber sawn from a batch of logs. An example of simulated conversion results is shown in Figure 2 and an extract from the summary for the results is shown in Figure 3.

### *Design*

In the design of a sawmill it is important to test the effect of a particular machinery configuration on yield and through-put for the anticipated log supply and markets. Simulation runs for a known log top diameter distribution will give the yield of main product battens and the number of boards and battens which require edging and cross-cutting. When these data are combined with the rate at which logs are processed by the primary machine the viability of a complete sawmill machinery

| Top diameter (cm) | Cant size (batten width) | Batten thickness (mm) | Number of battens | Total yield (% of log vol.) | Batten volume (m$^3$) | Full length batten yield (% of log vol) | Thickness of outer batten and boards (mm) | | Volume of edged and cross-cut battens (m$^3$) | Volume of boards (mm) | Yield of edged and cross-cut battens (% of log vol.) | Yield of boards (% of log vol.) |
|---|---|---|---|---|---|---|---|---|---|---|---|---|
| 18.4 | 100 | 47 | 3 | 54.59 | 0.069 | 42.86 | 47 | 19 | 0.000 | 0.019 | 0.00 | 11.73 |
| 18.4 | 125 | 47 | 2 | 56.11 | 0.057 | 35.72 | 47 | 19 | 0.000 | 0.033 | 0.00 | 20.40 |
| 18.4 | 150 | 47 | 2 | 54.59 | 0.069 | 42.86 | 47 | 19 | 0.000 | 0.019 | 0.00 | 11.73 |
| 18.4 | 175 | 47 | 1 | 55.87 | 0.040 | 25.00 | 47 | 19 | 0.000 | 0.050 | 0.00 | 30.86 |
| 18.6 | 100 | 47 | 3 | 54.43 | 0.069 | 42.04 | 47 | 19 | 0.000 | 0.020 | 0.00 | 12.39 |
| 18.6 | 125 | 47 | 3 | 61.29 | 0.086 | 52.55 | 47 | 19 | 0.000 | 0.014 | 0.00 | 8.74 |
| 18.6 | 150 | 47 | 2 | 54.43 | 0.069 | 42.04 | 47 | 19 | 0.000 | 0.020 | 0.00 | 12.39 |
| 18.6 | 175 | 47 | 1 | 55.68 | 0.040 | 24.52 | 47 | 19 | 0.000 | 0.051 | 0.00 | 31.16 |
| 18.8 | 100 | 47 | 3 | 53.39 | 0.069 | 41.24 | 47 | 19 | 0.000 | 0.020 | 0.00 | 12.16 |
| 18.8 | 125 | 47 | 3 | 60.79 | 0.086 | 51.55 | 47 | 19 | 0.000 | 0.015 | 0.00 | 9.24 |
| 18.8 | 150 | 47 | 2 | 54.26 | 0.069 | 41.24 | 47 | 19 | 0.000 | 0.022 | 0.00 | 13.02 |
| 18.8 | 175 | 47 | 1 | 54.62 | 0.040 | 24.06 | 47 | 19 | 0.000 | 0.051 | 0.00 | 38.56 |

FIGURE 2. Extract from a summary of simulation results for a Chipper centre and 'quad' band saw, primary breakdown configuration cutting 47mm thick battens and 19mm boards.

Top Dia. 18.8 Length 5.1m Volume 0.17 Highest yield 53.39%
CANT Battens 3 off 47 × 100 Length 5.1m
FIRST PASS Side Product Per Side 19 × 125 × 4.5m
SECOND PASS Side Product Per Side
FIRST PASS 0 19 100 19 0
SECOND PASS 0 0 147 0 0
THIRD PASS or
RESAW Setting 47 47 47

Top Dia. 18.8 Length 5.1m Volume 0.17 Highest yield 60.79%
CANT Battens 3 off 47 × 125 Length 5.1m
FIRST PASS Side Product Per Side 19 × 95 × 4.5m
SECOND PASS Side Product Per Side
FIRST PASS 0 19 125 19 0
SECOND PASS 0 0 147 0 0
THIRD PASS or
RESAW Setting 47 47 47

Top Dia. 18.8 Length 5.1m Volume 0.17 Highest yield 54.26%
CANT Battens 2 off 47 × 150 Length 5.1m
FIRST PASS Side Product Per Side
SECOND PASS Side Product Per Side 19 × 125 × 4.8m
FIRST PASS 0 0 150 0 0
SECOND PASS 0 19 97 19 0
THIRD PASS or
RESAW Setting 47 47

FIGURE 3. Examples of simulated conversion results showing the settings and products for a Chipper center and 'quad' bandsaw primary breakdown machine.

configuration can be analysed.

*Maximising Yield*

The use of optimisation programs may vary between low and high investment sawmills. In the former the results of background simulation runs may be used simply as an aid to operator instruction. If top-diameter log-scaling is carried out, they could also be used to generate manual look-up tables which the operator can refer to in order to set his machine in the correct way for each log.

In higher investment sawmills, the current practice is to use a semi-automatic primary log breakdown configuration which comprises a chipper canter followed by a pair of band saws or 'quad' bandsaws. It is essential to know exactly what sizes are to be sawn from a particular log in order to set the chipper-canter. The width of timber left between the two chipped faces must be an exact multiple of the sawn sizes required and saw kerf. The internal control system of this configuration includes a memory which holds pre-selected preferred cutting patterns which are linked to a sawlog top diameter classification system. The preferred cutting pattern is accessed and the chipper canter and saws are set-up correctly when the log top diameter is input into the control system of the machinery.

*Log Classification Systems*

A computer optimisation program is the only way of accurately and quickly designing a log classification system and selecting the preferred cutting patterns to be put into the memory for the chipping and sawing configuration. There are two elements which should be considered in order to give an optimum conversion, within constraints, for a batch of logs:

a) The range of top diameter in each class.
b) The optimal conversion pattern to produce the required spread of sawn sizes from each class.

The use of simulation runs can lead to a finely tuned classification system. However, it is only sensible to use it if logs are being measured accurately as in high investment sawmills which usually automatically scan log diametes in at least one plane. Log classification systems can also be used in sorting in the log yard in order that a group of logs, all of the same classification (top diameter range), can be fed to a sawing configuration which is set-up to give the optimal conversion. This system maintains very high through-put by eliminating the need to set the saws for each log and, within the constraints imposed by the equipment used, optimises the yield.

MODELS OF SAWLOG SHAPE

*Truncated Cone Approximation*

The techniques described use optimisation programs which approximately sawlogs to truncated cones. However, these models are limited because they allow only uniform cross-sections, taper, log length, saw kerf and wane tolerance to be considered. In the UK, logs are very rarely perfectly tapered and straight, they often have irregular cross-sections and can be bent in one or more planes. Therefore measuring systems and optimisation programs should be able to deal with true log shapes. Some improvement in optimisation of the yield from a batch of irregular shaped logs can be gained by continuously measuring the diameter in two planes at 90° to each other and linked to a common datum. This makes it possible to estimate

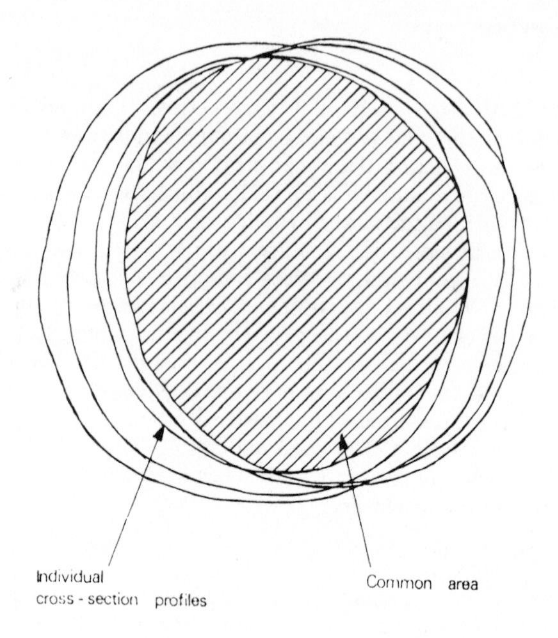


FIGURE 4. Log common area.

the angular orientation of the plane of maximum bow and also the common area diameter (e.g. Figure 4). This can then be used for log sorting or to access a stylised conversion of the common area which has been previously derived by using optimisation programs based on truncated cones. More advanced optimisation programs allow the truncated cone to be uniformly bent in one plane or in two planes at 90° to each other. However, these programs only allow the effect of bow and irregular cross-section on yield to be estimated and therefore do not predict the optimum yield.

*True Log Shape Model*
The only way of accurately predicting the yield from irregular logs is to simulate their true shape. Towards this end PRL has developed and tested a unique computer optimisation model. It differs from those based on truncated cones in that it is able to consider logs or irregular cross-sectional shapes and is not dependent on simplifying assumptions related to log bow. It also allows irregular longitudinal wane to be considered and can simulate accurately the conversion of slabs to random length and width boards. The PRL model was developed on a main frame computer some years ago and although it has been used to develop stylised cutting patterns its main use has been in research. However, recently a version of the program has been modified to run on an Acorn BBC micro-computer and has been used to test the influence on yield of a North American sawmilling configuration. The optimisation program allows for single and double taper sawing, asymmetric positioning of the main cant, precise wane allowances on sawn timber and the conversion of slabs to random length and width sawn boards.

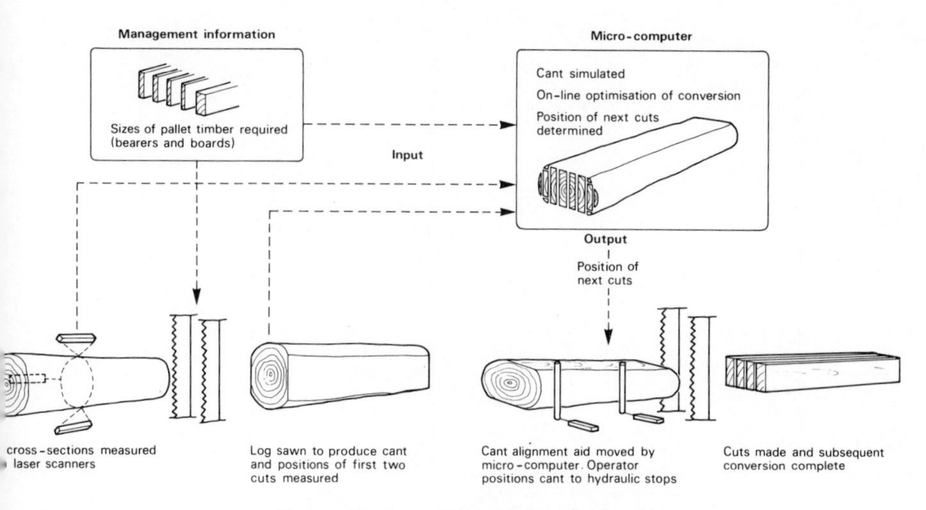


FIGURE 5. General layout of LOCUS.

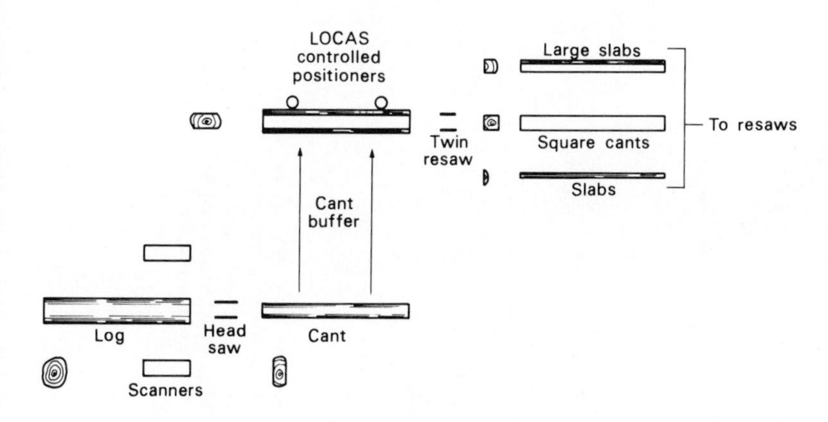


FIGURE 6. Diagram showing LOCUS installation.

'ON-LINE' OPTIMISING SYSTEM

Although the use of background simulation programs has improved the conversion yield in many sawmills the individual logs themselves are not necessarily being processed to produce their optimum yield. This can only be achieved consistently, throughout a working day, by using an 'on-line' decision making process (Maun, 1977) and control of conversion. Towards this end PRL have developed LOCAS. This is unique for the following reasons:

i) it deals with actual log shapes
ii) it determines optimal conversion of these shapes in real time
iii) it can be installed in a 'retro-fit overlay' basis into existing sawmills, as well as in new mills.

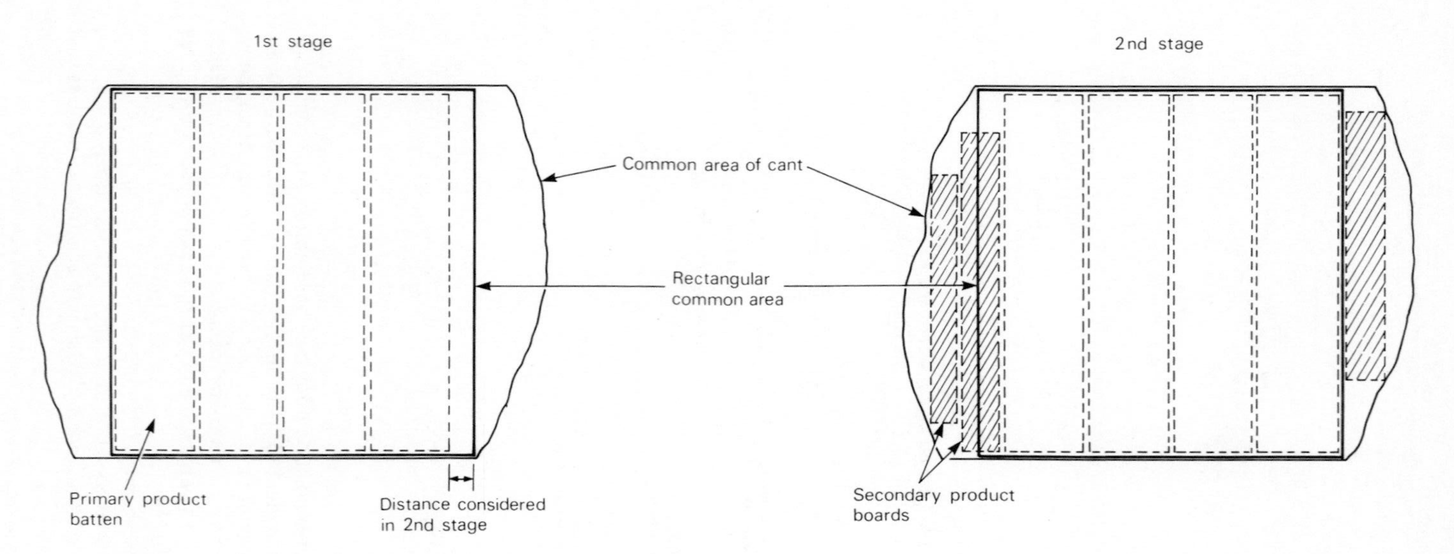


FIGURE 7. Two-stage optimisation showing only one size of primary product.
1st stage: Four primary product battens can be cut from the rectangular common area but, positioned like this, the resulting slabs would yield only two boards of secondary product.
2nd stage: Using the distance available across the rectangular common area, LOCUS repositions the battens so that slab's are produced which will yield more secondary product.

As each log passes through the headsaw, laser scanners obtain the cross-sectional profiles of the log at short intervals along its length, thus enabling taper, bow and irregular cross-sections to be considered in the optimisation procedure. The log shape information is fed to a high speed micro-computer which evaluates the alternative ways of carrying out the conversion and selects the optimum, according to the priorities for which it is programmed. The micro-computer then provides instructions down line for further conversion of the log, see Figure 5. The equipment built by PRL, for experimental and demonstration purposes, is designed for use with a double-band headrig and the optimisation is carried out on the cant which emerges from this machine, see Figure 6. The scanners are placed on the in-feed path to the saw and the simulation takes account of the position of the primary saw cuts. As a cant alignment aid, the experimental system uses two hydraulic stops which are controlled by the computer, see Figure 5. The micro-computer optimisation (Maun, 1980), which is restricted to the sizes of sawn timber required, is carried out in two stages. The first stage considers the conversion of full length main products from the area which is common to all cross-sections—the common area. This conversion rarely uses the entire width of the cant face available, see Figure 7, and the cuts can be positioned in a number of places without altering the yield of the full length product. This available movement is used in the second stage of optimisation, which determines the exact position of the common area cuts in order to maximise the total yield.

In a controlled test (Maun, 1983) in a sawmill, lasting several months, LOCAS demonstrated conclusively that it offers real advantages over manual control of conversion. It gave more control of the sawing of the main product and increased the amount of sawn timber by 7.5 percent. This was not achieved at the expense of the secondary products because the total yield was increased by 5 percent. It is estimated that the resulting increase in income would pay back the cost of the equipment and its installation in a medium investment sawmill (20 000 cubic metres of sawn timber out per year) within 2 years. The pay back time for a large investment mill (over 50 000 cubic metres out) will be shorter.

### *Other Application*

The basic components of LOCAS (Smithies, 1983)—the accurate log measuring device coupled with software to simulate conversion—could be used to control conversion in both existing and new sawmills of considerably different design. The PRL experimental equipment is just one example of how LOCAS could be implemented, others are shown in Figures 8, 9 and 10. In addition to using LOCAS purely for the optimisation and alignment of cants, the scanners and computer could be used as a sophisticated sorting system in the log yard. Logs could be scanned, classified according to their common area and the best rotational orientation of each one determined by the computer which could control a spray head producing a coloured line on the end of the log. The saw operator would then orientate this mark to a vertical position as the log entered the production line, ensuring that the log was in its best angular orientation for the optimum yield of sawn timber.

## CONCLUSION

Modern high investment sawmills are capable of processing large quantities of logs but the cutting pattern and orientation must be correctly determined for each log.

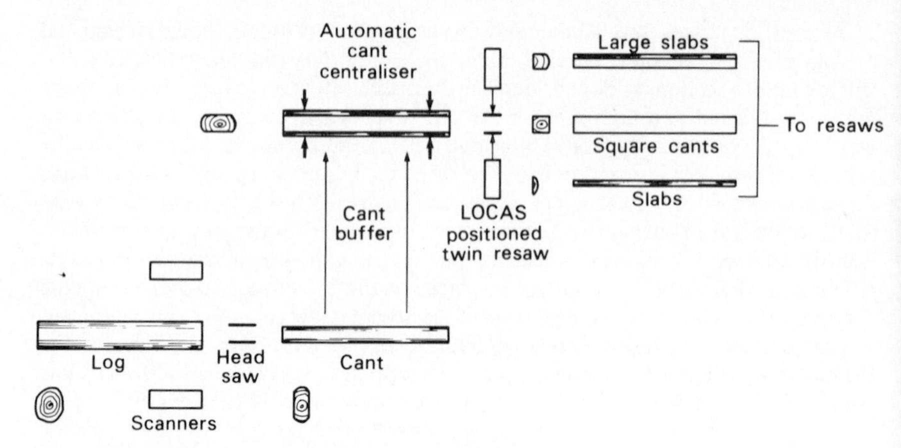


FIGURE 8. Twin bandsaw line with LOCUS controlled cant saw.
1. Logs are measured as they approach the headsaw.
2. Cants are automatically centralised at the second twin bandsaw; LOCUS controls the position of the saws and their distance apart.
3. Conversion is completed on further resaws.

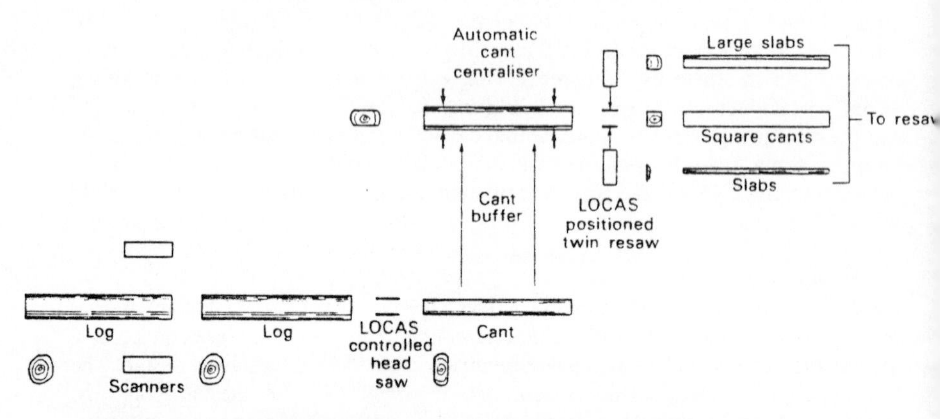


FIGURE 9. Twin bandsaw line with LOCUS controlled headsaw and cant saw.
1. Logs are measured on the extended infeed conveyor.
2. LOCUS sets the position of both headsaw cuts.
3. Cants are automatically centralised at the second twin bandsaw; LOCUS controls the position of the saws and their distance apart.
4. Conversion is completed on further resaws.

With the use of chipping heads there is no second chance to improve the yield of 'butchered logs' by reclaiming material from the slabs as there are no slabs, only chips.

It is not difficult to run a modern mill at full capacity and still make a loss because the conversion yield is too low, thus a 1 percent loss in conversion yield can mean an

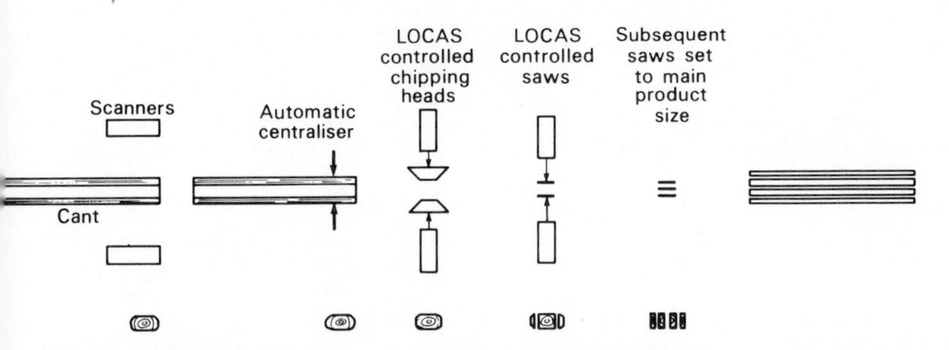


FIGURE 10. Reducer bandsaw line with LOCUS control of cant conversion.
1. Cants are measured prior to entry to the saw.
2. LOCUS controls the position of the chipping heads and of the saws of the first twin bandsaw so that the yield of boards is maximised.
3. Conversion is completed on subsequent saws which are set to the required dimensions for the main product.

extra 1 500m$^3$ of logs have to be processed to maintain the same annual return on investment.

Hence it is vital to maximise the conversion yield as well as maintaining the conversion rate.

Sawmills can only achieve these two aims and adapt to changing markets by the use of computers. The use of background optimisation programs help but, 'on-line' systems, such as LOCAS, offer more advantages and will become increasingly more important in the next decade.

REFERENCES

Maun, K.W. (1977) An economically viable computer-aided conversion system for British sawmills. BRE Information sheet IS 19/77.

Maun, K.W. (1980) The concept of LOCAS. Paper presented at the PRL seminar 'Best yield from British sawlogs' held at Nairn, November 1980.

Maun, K.W. (1983) LOCAS The Assessment. Paper presented at PRL seminar held at Lampeter March 1983.

Smithies, J.N. (1983) LOCAS The Installation. Paper presented at PRL seminar held at Lampeter March 1983.

# COMPUTERS IN THE BRITISH COLUMBIA SAWMILL INDUSTRY

GEORGE H. STICKNEY

*Hipp Engineering Ltd, Vancouver, British Columbia*

SUMMARY

This paper reviews some applications of computers in the sawmill industry in British Columbia. The paper describes briefly the structure of the industry and the benefits derived from computers.

## INTRODUCTION

Hipp Engineering is a Vancouver-based company, established in 1973, which provides consulting engineering, detail design and project management services to the forest products industries in Canada and the USA. The company has extended its services in the international markets through a cooperative agreement with EFG Consultancy Division in the UK. The sawmilling experience among the technical staff includes western and eastern Canada, the southern USA, New Zealand, Australia and South America, ranging from large to modest scale mills.

The author's experience covers 25 years as a professional engineer in Canada's forest industry. This includes direct involvement in all aspects of production, maintenance, research, new construction, and consulting in sawmilling and plywood and their interface with logging, pulp and sales.

### *Introduction to the British Columbia sawmill industry*

When discussing the sawmilling industry in B.C., recognition must be given to the influence of the well-established pulp industry in B.C. as well as the fact that the industry produces predominantly for export markets.

There are well over 300 sawmills in B.C. and these may be divided into two distinct sectors: the interior sawmilling groups and the coastal sawmilling operations. The divisions between these two groups are reflected in several major factors, these being:

- different mill designs and capacities,
- different average sizes, with mills being smaller in the interior,
- coastal mills are much more closely integrated with the pulp industry.

The interior sawmill industry mainly processes spruce, pine and fir with some cedar and Douglas fir. The logs generally are small diameter, from 10 to 50cm. They produce three products, dimension lumber, veneer and pulp chips. Logs are delivered by truck; lumber is shipped largely by rail to either the US or export ports and the chips are transported by large chip trucks to the interior pulp mills or by rail to the coastal pulp mills or chip export facilities. Interior sawmills produce a large amount of kiln-dried lumber for the US market whereas many of the coastal mills ship green or air-dried timber for export.

The coastal sawmills are processing larger diameter sawlogs and are usually more closely integrated with the pulp mills. The mills are generally located on the coast and process mainly Douglas fir, hemlock, balsam fir and western red cedar for export markets. The mills produce principally dimension lumber, chips for the pulp mills and in numerous cases sawdust for pulping. Being closely integrated with the pulp industry, many of the coast mills operate virtually as wood preparation plants for the pulp mills.

## COMPUTER APPLICATIONS

### *General*

In the B.C. sawmilling industry the use of computers is expanding. A large proportion of the interior sawmills producing over 200m$^3$ per shift or 100 000m$^3$ per year on a two-shift basis utilise computers to increase overall recovery and efficiency. Around half the mills scan the incoming logs to record incoming volumes and select cutting patterns on the headrigs.

A sawmill is basically a series of processing units linked by material transfer equipment. In operating a sawmill, operators have in the past been required to make a great many decisions as the logs or sawn-timber pieces are fed through their machines. The possible misjudgement or error in these decisions can have a profound impact on the profitability of the company. In a modern mill producing some 100 000m$^3$ per year, a one percent loss in sawn-timber production through errors in judgement costs about $35 000 in profit.

The sawmills use the computer-scanner to improve productivity in terms of through-put and value recovery, since the system can reduce the incidence of operator error. The modern sawmill is a blend of human and computerised operations. The feeding and positioning of logs and sawn pieces prior to the sawing units is best handled by human operation, whereas the scaling of logs and selection of cutting patterns is more accurate and faster by computer.

Computers are used throughout the mills in such functions as:

- log measuring, sorting and cross-cutting,
- log scanning and data recording, selecting of cutting patterns and setting of saws,
- optimising edging and trimming,
- optimising the rate of lumber drying relative to quality,
- stock inventory control, order control, shipping recording and invoicing,
- overall mill management systems including market requirements, wood characteristics and all mill functions,
- control of processing and materials handling equipment, particularly the timing and sequencing of operations.

### *Log preparation and cross-cutting* (Figure 1)

An example of the utilisation of computers in optimising log cross-cutting is the system installed by Crown Zellerbach at their Elk Falls wood processing complex on Vancouver Island. Being integrated with pulping facilities, the objective was to eliminate waste material at the first opportunity. The key to the selection of truly waste material is the scanner computer system. Logs are fed to the mill in long lengths one at a time and debarked in long lengths. A defect saw is located ahead of the debarker for the removal of severe defects. Pieces containing defects are removed to the woodroom for chipping. Debarked logs are scanned and measured for diameter,

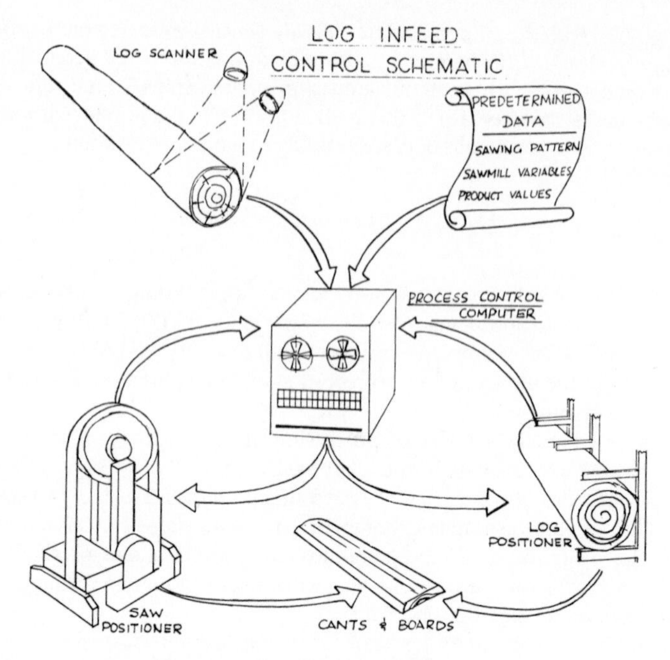


FIGURE 1. Log infeed control schematic.

length and taper. They are then automatically directed to one of two lines where the logs are cross-cut. The data fed from the scanner ascertain when the in-feed to the cut-off saw station is available for the log and transmits a recommended log cutting pattern to the cut-off saw operator. The impact of optimum log cutting has been shown to result in a possible ten percent increase in sawn lumber.

The objective in computerised log-cutting is to trim the logs to accurate lengths while achieving log length selections giving maximum profit per cubic metre of logs.

In this system the actual positioning of the logs for cross-cutting has been left to the operator who uses positive mechanical log stops. It was felt that computer control was subject to possible error due to log slippage on conveyors and that the operator would be required anyway, so little was to be gained by full computer control.

*Sawing*

The most common and well-known application of computers is log scanning, selection of cutting patterns and control patterns and control and setting of chipper heads and saws.

Computerised saw control systems allow the operators to achieve accurate sawing at a high linear throughput while optimising cutting patterns in terms of sawn volume and economic recovery. The human operator is only able to utilise a relatively small number of sawing patterns. The computer can have over 2 000 predetermined sawing solutions stored in the memory which can be selected according to log diameter and taper dimensions. The computer will then set the chipping sawing equipment while the operator concentrates on positioning the log to best advantage and maintaining a

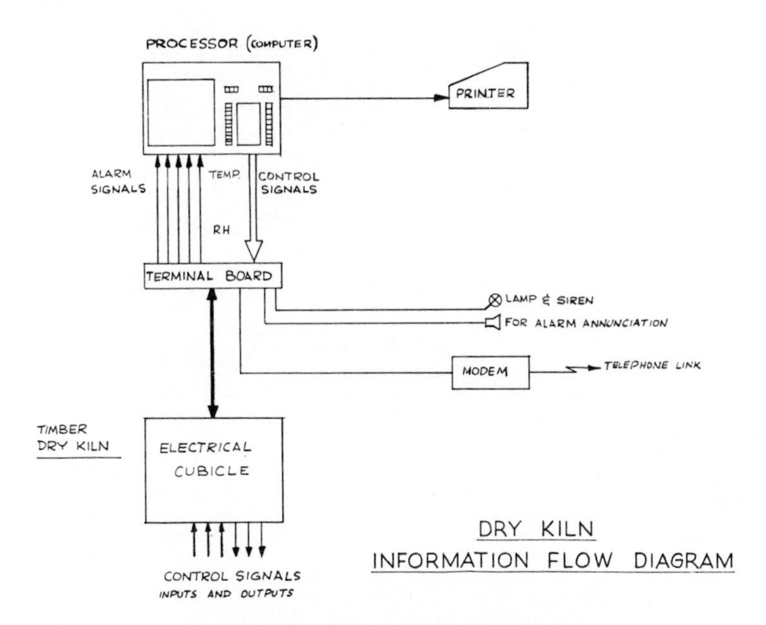


FIGURE 2. Dry kiln information flow diagram.

smooth flow of logs. The same principle applies to the subsequent sawing of cants.

*Optimiser edging and trimming*

The use of scanners to control the edging and trimming process is becoming more practical. Essentially the scanner 'reads' the board and calculates its shape particularly relative to wane on the edges and tapered ends. The best full board that could result from edging and/or trimming this piece can then be calculated according to the essential program parameters.

At present edging and trimming are two distinct operations, each requiring its own scanner and software. Both scanner systems are generally known as optimisers.

Final grading and sorting is usually subsequent to the trimming operation and here again optimiser decisions are visually assessed by parameters to yield decisions but scanners and other related equipment have limited capabilities in measuring the characteristics that contribute towards the final lumber grade,

*Dry kilns* (Figure 2)

Another interesting application of computers which has proven very successful is the kiln drying program. A B.C. supplier has sold over 50 of these systems in the Province, at a price of about $25 000 each.

Kilns traditionally work on what is termed an 'open-loop system.' The operator selects a schedule based on prior experience from a possible list of more than 100 and the wood is dried at the selected temperature cycle. The new computer application is a 'closed-loop system' in that information from the kilns is relayed back to the computer, which can then adjust the drying time accordingly. The computerised kiln

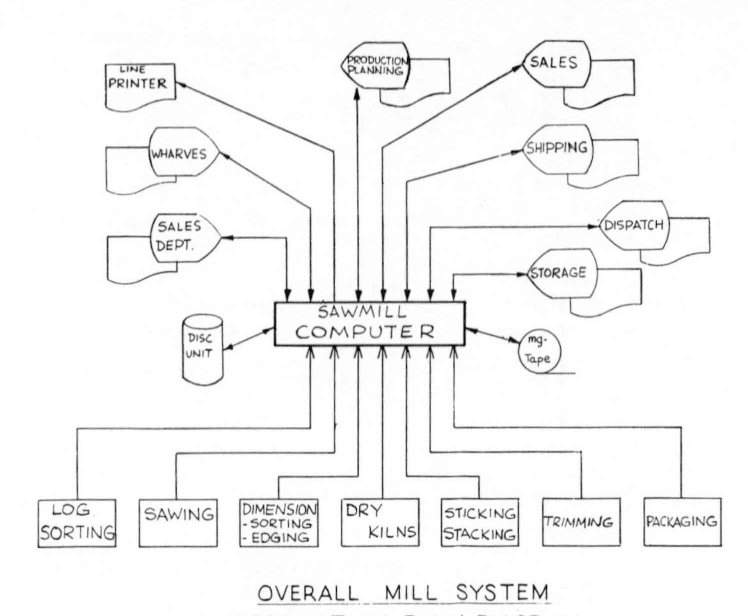


FIGURE 3. Overall mill system information flow diagram.

system has monitors that read the temperature and moisture levels in the wood, revealing the status of the wood drying process. The lumber in the kiln thus undergoes a drying cycle monitored and controlled by its own specific drying rate.

*Overall management systems* (Figure 3)
Computers can assist management in planning changes to mill flow and calculate the potential benefits of changing the product mix within the mill. These analyses are more complicated than evaluating just one mill sub-system because of the interactions between each piece of machinery. Relative capacities must be considered. Computer programs which analyse whole-mill situations produce only approximations. By integrating sub-systems to develop the whole mill, the actual mill performance can be determined.

OBSERVATIONS
The newer generations of micro-computers are becoming as powerful as the larger computers of ten years ago. A desk-top system of the 1980s can offer the computing resources which formerly occupied several cabinets of equipment. Before anyone rushes to pursue one of these systems, though, it must be noted that powerful computer equipment alone does not solve the problems. It is the software or computer programs which must be considered first, not the equipment. Too many users have purchased powerful 'state-of-the-art' computer systems only to find that there is no suitable software or that the programs do not include the functions they thought they were getting.

Large linear programs have been developed which examine combinations of product flow, piece counts and breakdown possibilities. These programs attempt to select an optimum solution. However, even with the complexities of all factors considered these programs still require the judgement of an experienced sawmiller. No mill runs evenly throughout a shift without surges or backups, receives ideally tapered logs, or avoids mechanical and electrical problems. Recent advances in programming are starting to include provision for sweep and defective logs and are accounting for uneven product flow. The most important thing about these computer planning tools is that the user recognises their limitations and their benefits.

The trend is now towards localised computer station control and away from main-frame control throughout the sawmill. Despite the power of the computer the intelligence of the analyses and the relevance of the results lie in the logic of the programs. The key to the usefulness of planning programs is in the ability to evaluate and rate alternative solutions. Mill profits are dependent on good management decisions and the right decision requires the right information. It is advisable to utilise the services of a computer professional when selecting the hardware and software package. This is one critical area in which computers cannot be regarded as 'Black Boxes' where you merely observe the result to judge performance.

With computer planning the results are not as easily verified as with a piece of breakdown machinery where you can watch the wood entering and leaving the system. The logic behind analysis programs is hidden from the user and must be carefully investigated by someone with computer training to be certain that they are actually performing the calculations you expect.

If you are not an expert in this type of analysis it is advisable to retain the services of an independent professional. If the computer system vendor really has the product he claims, then it will stand up to close scrutiny.

If a mill understands the relationship between market demands, products mix, mill production capacities and operating costs then a person skilled in the use of computer models can build a complete economic model of the company. This model can then be used by management to determine the outcome of the particular marketing or production strategy.

A fully developed economic model becomes a profit-planning tool for the future because it permits management to test how their cash position would change under different economic conditions. As the last several years have amply illustrated, being able to analyse the effects of economic change is essential to survive the difficult times and being able to maximise profits in good times.

If our respective industries are to maintain a competitive position with worldwide producers of wood products, then management must adopt computer techniques to assist in strategic planning.

# COMPUTERS IN MARKETING AND MARKET INTELLIGENCE IN THE TIMBER IMPORTERS MARKET

A.A. SINCLAIR

*'Braddan', Station Road, Windygates, Fife*

## INTRODUCTION

The purpose of this paper is to describe the experience of a medium-sized timber importing company based in Scotland, in its efforts to meet the marketing challenges of the eighties by the use of computers.

It attempts briefly to illustrate the factors influencing the author's choice of computer and implementation as well as describing the types of reports our system can currently produce or that are under development.

## BACKGROUND TO COMPUTER REQUIREMENTS

In 1978 the company, of which I am Finance Director, embarked on a programme of diversification and expansion. Within a short space of time it moved from almost exclusively softwood importing to the creation of a Do-It-Yourself (DIY) store and the launching of a roof-truss fabrication plant, both in Leven, the setting up of a satellite depots in Dundee and Glasgow, and the acquisition of a company based in Edinburgh. A garden centre was subsequently added to the 'DIY' store. Strategic marketing had dramatically increased sales of panel products and hardwoods.

The company's executives had clear ideas on the management information systems required to control marketing, sales, buying, production, stocks and finance. Annual, and later, five-year budgets were considered an essential discipline as was the prompt production of actual monthly results monitored against budget forecasts.

It became apparent that these information services could only be provided in a cost-effective manner by the use of computers. My interpretation of the company's detailed objectives for the introduction of computers is reproduced in Appendix I.

I hope it is clear from Appendix I that there were perceived tasks and problems before any decision was reached to acquire a computer.

## COMPUTER SELECTION (HORSES FOR COURSES)

### *Pitfalls and how to avoid them*

The most important point for any prospective first-time computer user is to take professional advice from the best source available. You should employ a consultant from sources such as:

your local University—Computer Studies Department;
the Management Consultancy Arm of your Auditors;
Trade Association; and so on.

Give the task to a professional who is objective and has *your* best commercial interests at heart.

For anyone who has not had in-depth computer training or experience the chances of failing to identify all the critical aspects of a prospective computer system are extremely high.

You will not understand the salesman's 'jargon', and frequently he will not understand how your business operates.

You will overlook engineering response times, critical printer speeds, time required to take security copies, even the financial security of the supplying organisation or its commitment to continue manufacture and more importantly, development of the product in which your are interested.

By and large, computer salesmen are a cynical breed, apparently motivated by commission rather than customer satisfaction. There are exceptions, of course, but only someone with specialist knowledge and skills can steer you in the right direction. The problem is to blend your requirements and cost limitations with what is available in the computer market.

The system we adopted may not be the best for all importers, let alone for other arms of the timber industry.

### *Acquisition and Installation*

How soon after placing an order can you expect some results? In a larger system such as a mini-computer with say 30 megabytes of back-up storage capacity, five visual display (VDU) screens and two printers, covering a wide range of activities, our experience was that the period from order to delivery was 4-6 weeks, barely long enough to complete the necessary electrical work.

Thereafter, because we were using programs already tested and developed, we were able to start entering basic names and addresses into the system within two working days of delivery.

We had recruited temporary staff to cover the period when our old and new systems would run in parallel. Because the operating system was sophisticated, but the programs themselves easy to understand, we were able to have several people carrying out the same task simultaneously.

Consequently, a new set of programs were implemented on average every six to eight weeks.

I cannot over-emphasise the need for the parallel running of old and new systems. Every suite of programs must be individually tested by parallel running and, in my opinion, for routines such as wages, salaries, sales/purchase ledger and trial balances, a minimum of two end-of-month comparisons is essential—hence the eight week gap between the setting up of each new routine.

## COMPUTER AIDED MARKETING INTELLIGENCE

Broadly market intelligence is information to assist an organisation to match its strengths to the demand of its markets. Of particular relevance in our case is the fact that, being a privately owned company, we rely heavily on retained profits to fund expansion in working capital. Therefore the forecasting of profit and working capital is deemed to be of paramount importance, as is actual performance measured monthly against forecasts. Our main computer system (a Nixdorf Model 8870/35 mini-computer) processes normal accounting transactions but captures data for marketing analysis at the same time.

Thus we know how much of each product we have sold to each customer and at what gross margins both for the month and year to date. We can tell which division of the company made the sale or summarise sales and gross margin per product by a specific geographical area (it could be per salesman if so desired). We can assess softwood sales by qualities (e.g. Appendix II) and measure sales volume, sales value and sales value per cubic metre distinguishing between sales with and without added-value opportunities. The total average price per cubic metre over all softwood products is a valuable market barometer particularly if measured in relation to changes in cost value per $m^3$ (See Appendix III). This has the same basic matrix as Appendix II but shows cost value, sales value and gross profit per softwood product group and in total. Total gross profit over all softwood groups is another valuable indicator.

Due to the very low levels of fluctuation in operating costs from one month to another in our softwood mill, the monthly operating profit can be very accurately forecast as soon as the request shown in Appendix III is available. We know the outcome for a given month four days after the end of the month. Under the previous system we would have waited for up to 10 days longer.

We have begun to tackle detailed stock control not only for 'DIY' products, but for softwood. We estimate this will take around a year to plan and implement, but the programs to handle the work are already available on both our own and other hardware in the form of fully developed packages.

These products of the mini-computer are passed directly to marketing and mill management, or to our 'DIY' manager unaltered, but not all output from the mini-computer is of the required standard.

The present facility on the mini-computer to produce management accounts is not entirely satisfactory purely in terms of layout and presentation. Therefore we pass the sophisticated trial balance produced on the mini-computer through a micro-computer (an IBM Personal Computer using Lotus software) to produce a wide range of management accounts (e.g. Appendix IV).

There are seven similar sets plus a consolidated statement for the total company. The layout is almost infinitely variable and can be changed in a matter of a few minutes, which was not the case with similar reports produced by the mini-computer where the layout was relatively fixed. The micro-computer also maintains all our budget spreadsheets and allows total annual budgets to be analysed monthly on a wide range of bases to allow for seasonal weightings, holidays, price increases and so on.

Over the last two years we have been developing a financial model of the company on the micro-computer. The model is constructed on a spreadsheet program which uses cells, each with a grid reference, thus providing a sheet of electronic graph paper. Each cell can be purely a description, a value which is keyed in or a value arrived at by evaluating a formula (e.g. gross profit = sales value minus cost of sales value). Everything that we do on our micro-computer is based on the same program and financial modelling is therefore only a variation of our budget spreadsheets by consolidating the trading forecasts of our individual profit centres whilst using sales to predict debtor levels and cost of sales to compute implied stock balances.

Consequently if the model is correctly constructed it will measure the effect (where appropriate) of changes in sales levels, gross margins, operating costs, in any of our seven profit centres and show the resultant values for net profit, capital employed, shareholders funds and bank position.

This allows a wide range of budget options to be evaluated in a few minutes, a task which would otherwise take hundreds of man-hours.

## CONCLUSION

I believe that whilst we will continually be refining and improving our systems, we have nonetheless already developed a series of computer-produced reports which serve as effective market 'sensors' allowing us to detect quickly, and to quantify the effect of, changes in our markets. We know where our profits are earned and what capital is required to service each operation, consequently as each year goes by we become more confident of our ability to predict the future and to be able to take positive action promptly, to ensure that we achieve our stated objectives.

## APPENDIX I

### DETAILED OBJECTIVES

1. *Budgeting*

a) Produce an annual budget analysed by months (taking into account seasonal factors) for each of 7 profit centres (P.C.'s). Sales value and gross profit to be analysed by cash credit sales and by product groups within each sales type

b) Analyse capital employed by P.C.'s as above and measure the return thereon

c) Consolidate P.C.'s budgets to produce forecast company performance by month and year to date

d) Forecast balance sheet at half-year and year-end

e) Produce a financial model of the company capable of evaluating the effect of changing any of the component parts of the company budget to show:

i) Changed net profit and express as a return on shareholder funds (at the beginning of the accounting year)

ii) A revised balance sheet and implied funding requirements.

2. *Marketing and associated data*

a) Produce analysis of sales and the gross profit thereon by product within customer (month and year to date)

b) Produce totals of (2) (a) by four geographical sales areas

c) Analyse softwood sales to show $m^3$ by principal qualities distinguishing between merchanted and processed sales. See Appendix II

d) Analyse gross profit on softwood as per (2) (c). See Appendix III

e)Analyse sales depot/DIY sales and gross profit.

3. *Stock control*—To be tackled in two distinct phases

*Phase I.* To produce stock control data for the DIY Store (5 000 items) to meet specific objectives:

a) To improve stock turnover by identifying slow-moving items.

b) To eliminate the need for costly physical stock counts by creating accurate stock records per item, test checked by perpetual inventory techniques.

c) To analyse gross profit by stock items and produce a league table of item by item profitability.

*Phase II.* Automate the softwood stock records to identify stocks by location, quality, end section and length specification.

Calculate moving average cost value (ACV) and evaluate $m^3$ times ACV by quality.

Phase II would also involve the creation at the sawmill of yard orders and advice notes for deliveries. It is envisaged that advice notes and invoices will be produced simultaneously for pre-quoted softwood items, with descriptions, selling and cost prices being picked up automatically from the stock records.

4. *Accounting work*

Process:

Sales Invoice
Purchase Invoices
Wages
Salaries
Giro Transfers
Statements
Remittance Advice/Cheques
Journals
Stock Records
Credit Control Reports.

Processing to be integrated so that e.g.:—the production of a sales invoice would automatically update:

Sales Ledger
V.A.T.
Nominal Ledger
Stock Records

and create Journals etc.

Produce monthly management accounts measuring profitability and return on capital by profit centre both for the month and year to date. See Appendix IV.

Consolidate to produce company result and show profit after tax as a return on shareholders funds.

Produce a monthly balance sheet.

Management accounts are to be published not later than the tenth working day after the month-end to which they relate.

*Volumes to be processed (approximately)*

| | |
|---|---|
| Sales Invoices | 2500 per month |
| Customer Accounts | 600 per month |
| Purchase Invoices | 1200 per month |
| Supplier Accounts | 200 per month |
| Stock Records | 7000 Items |
| Trial Balance | 600 Accounts |
| Wages | 100 per week |
| Salaries | 50 per month |
| Journals (Manual) | 100 per month |
| Cheques/Remittances | 200 per month |
| Sales Statements | 500 per month |

## APPENDIX II

PRODUCT ANALYSIS

| Description | | Current | | | Year to Date | | |
|---|---|---|---|---|---|---|---|
| | | Volume | Value | Av. Price | Volume | Value | Av. Price |
| American | Merchanted | 30 | 11083 | 363 | 129 | 47035 | 362 |
| Softwood | Processed | 12 | 6337 | 516 | 75 | 36060 | 475 |
| Total | | 42 | 17420 | 407 | 205 | 83095 | 404 |
| U/S White | Merchanted | | | | | | |
| | Processed | | | | | | |
| Total | | | | | | | |

## APPENDIX III

PRODUCT PROFITABILITY

| Description | | Current | | | Year to Date | | |
|---|---|---|---|---|---|---|---|
| | | Cost | Value | Profit | Cost | Value | Profit |
| American | Merchanted | 8971 | 11083 | 2111 | 37003 | 47035 | 10031 |
| Softwood | Processed | 3730 | 6337 | 2607 | 21660 | 36060 | 14400 |
| Total | | 12701 | 17420 | 4718 | 58663 | 83095 | 24432 |
| U/S White | Merchanted | | | | | | |
| | Processed | | | | | | |
| Total | | | | | | | |

APPENDIX IV

EAST DEPOT

*Trading Statement as at 31/8/84*

| August | | | Year to Date | |
|---|---|---|---|---|
| Actual | Budget | | Actual | Budget |
| | | Sales (Cash) | | |
| | | Sales (Credit) | | |
| | | Total Sales (Ext) | | |
| | | Sales (Internal) | | |
| | | Total Sales | | |
| | | less cost of sale | | |
| | | Gross Profit | | |
| | | Gross Margin | | |
| | | Operating Costs | | |
| | | Wages & Salaries | | |
| | | NI, Hol Pay & Pension | | |
| | | Electricity | | |
| | | Rates | | |
| | | Telephones | | |
| | | Print & Stationery | | |
| | | Repairs and Upkeep | | |
| | | Car Expenses | | |
| | | Depreciation | | |
| | | Insurance | | |
| | | Int. Handling | | |
| | | Deliveries | | |
| | | Advertising | | |
| | | General Charges | | |
| | | Bad Debts | | |
| | | TOTAL OPER.COSTS | | |
| | | CONTRIBUTION | | |
| | | % SALES | | |

RETURN ON CAPITAL EMPLOYED

| | Actual | Average | Budget | Variance |
|---|---|---|---|---|
| Fixed Assets | | | | |
| Stock (ave) | | | | |
| Debtors (ave) | | | | |
| less Creditors | | | | |
| Net Cap. Emp. 12 months | | | | |
| ANNUALISED CONT. | | | | |
| RETURN % | | | | |

# PART VI

# FUTURE DEVELOPMENTS IN RESEARCH AND MANAGEMENT

# FOREST RESEARCH AND THE COMPUTER

J.N.R. JEFFERS

*Institute of Terrestrial Ecology, Merlewood Research Station, Grange-over-Sands, Cumbria LA11 6JU*

SUMMARY

Developments in the transition from manual calculation to the present widespread use of computers in forest research are reviewed. The introduction of computers has greatly reduced the time required for routine calculations. Access to data has become easier, allowing integration of research information orginating from many sources. Precise communication between scientists has been facilitated. Computer-based simulation models have shown considerable promise in linking the findings of the researcher with the needs of the resource manager or administrator. Further expansion in the role of computers is anticipated, particularly with the advent of expert systems offering intelligent assistance to the user.

INTRODUCTION

Few, if any, of today's undergraduate students will be able to envisage the difficulties of forest research before the electronic computer became readily available, even though that time was only 30 years ago. I began my own research career in the Forestry Commission's Permanent Sample Plot Party in 1946, and, at that time, we did not have even the simplest of calculating machines.

For example, the first basal-area tables used in the Forestry Commission were calculated by hand by three people independently so as to check the results. Later we acquired Brunsviga calculating machines and a Facit or two, when work began on the Forestry Commission volume tables, using the volume-basal area line of Hummel (1955). The first real breakthrough came when the Forestry Commission Research Branch purchased an electrically-driven calculating machine, a Marchant Calculator, primarily for the analysis of experimental and survey data which were then beginning to accumulate as forest research began to gain in momentum after the War. Before that, calculation depended largely on manual methods, aided by the Fuller Calculator—a large spiral slide rule—and sets of tables of squares, like Barlow's Tables.

Even with electrically-driven mechanical calculations, computation was slow and unreliable. The analysis of variance for a simple randomised block would take several hours, and would need to be checked independently. Multiple regression for a reasonably small data set would take several days, and would be limited to the simplest options, with no opportunity of testing various transformations of the variables. The first multivariate analysis took six weeks to complete, and was hampered by an undetected error in the matrix inversion, an operation which itself took several days. Calculating eigenvalues and eigenvectors was a particular agony, depending on an iterative procedure which never seemed to end. Anyone interested in the more numerate kinds of forest research needed to be very determined, and to have a touch of masochism as well!

The introduction of the electronic computer came as a revelation to those of us who were working in this field, and the first opportunities to use such machines came in the mid-1950s. A few of the early machines like the NPL DEUCE and the Ferranti Pegasus were installed in national research institutes. It was possible, with persistence, to get time on such machines if you were also prepared to learn the machine codes by which they were operated. In this way, some scientists were able to gain some useful experience which stood them in good stead when the machines became more readily available as they were installed in universities, and later in research institutes. The Forestry Commission Research Branch, for example, acquired its own computer—a Ferranti Sirius—in 1963, but, prior to that, used two computers at the nearby Royal Aircraft Establishment in Farnborough.

This paper reviews some of the major advances that have taken place in research through the introduction of computers, particularly in forestry, and looks ahead to new developments in computing which will greatly increase the potential of forest research to improve forest management and practice.

## CHANGES DUE TO THE COMPUTER AND THEIR ADVANTAGES

### *Speed of computation*

The most important change that we expected—and the one which originally prompted interest in the use of computers for forest research—sprang from the sheer speed of computation. Even the earliest computers, once programmed to perform the necessary tasks, greatly reduced the time taken to do routine calculations. Analyses of variance, complete with the printing of the results, were done in minutes rather than in hours. Multiple regression, or even multivariate analysis, became possible for the first time as a regular, even routine, form of analysis.

Since those early days, computers have themselves greatly increased the speed with which computations can be made, and unlocked a virtual Pandora's box of analytical techniques for the research worker. In the last ten years, for example, the speed of computers has increased by an order of magnitude roughly once every two years. There are very few clerical or arithmetic manipulations which are beyond the scope of the research worker in forestry today, even if he does not fully understand the logic of the calculation that he has performed through the medium of the computer and its guiding programs. We already live in an age in which computational ability has outstripped our own ability to reason and think logically, and everything points to an even greater increase in this ability well within the next decade.

### *Access to data*

However, speed of computation did not prove to be the most important reason for using computers in research. Before the computer, it was the accepted practice, indeed the dogma, that all data must be recorded in field or laboratory notebooks. In long-term projects like the Forestry Commission Permanent Sample Plots, there were elaborate sample plot files that were added to at each re-measurement. While undoubtedly an excellent discipline, recording data in this way effectively inhibited their subsequent use. Most of the effort required to produce the Forestry Commission volume and yield tables went into extracting the data from the files in which they had been recorded for safe-keeping.

In order to use the computer, data had to be made machine-readable. At first, this meant transferring data to punched cards or to punched tape, the two standard media for transmitting numbers to computer memories. Today, data are usually input to some form of magnetic device, often a floppy disk, or magnetic tape, or directly to a hard, but exchangeable, disk. Once made machine-readable, however, these data can be used many times, and in combination with other machine-readable data, leading to the concept of a data-base, which may be the property of a single individual or the joint property of a whole research organisation. Given proper data-base management, the computer now becomes a powerful medium for the exchange of information between individuals, and for the integration of research results from a whole range of research scientists from different disciplines.

Collaborative projects of all kinds become possible when access to the data collected can be made freely available to the scientists working on the project. Each scientist adds his or her own data to the shared data-base, and can, in turn, make use of the other data to widen the range of hypotheses that can be tested. This improved access to a data-base, while made possible by the speed with which computers work, has turned out to be more important than the speed with which any single computation like, say, matrix inversion can be performed.

### *Algorithms and algorithmic languages*

Despite the importance of the two aspects outlined above, there is a third advantage that outstrips both, and, strangely, this flows directly from what, at first, seemed to be the major disadvantage of computers. In order to use a computer at all, it has to be programmed, thus there has to be a completely unambiguous set of instructions, or program, guiding the computer through all of the logical and arithmetical operations necessary for the correct completion of the computation. The early programs had to be written in machine codes which were specific to each make of computer, and sometimes even to individual machines. Gradually, however, easier ways of programming computers were developed, first as autocodes and then as high-level languages like Fortran, Algol and BASIC. Programming became easier while remaining exact, in the sense that the programs themselves are still unambiguous sets of instructions. The high-level languages also made these instructions easier to read by human beings as well as by computers, so that, again for the first time, we have detailed and very precise descriptions of how to perform logical and arithmetic operations on symbols (numbers and letters of the alphabet) to achieve some stated end. We call these sets of instructions algorithms.

The algorithms and the algorithmic languages were made necessary by the computer, but in a very real sense they are more important than the computer. We can communicate our ideas more readily and more effectively through algorithms, preferably with the aid of the computer, than by any other method, and much of today's communication between numerate scientists is through such algorithms.

## ACHIEVEMENTS TO DATE

What have we done with this new and revolutionary tool since computers first became available for forest research in the mid-1950s? Has the promise of this new tool been fulfilled, or has the change been less revolutionary than was expected? In general, the change has indeed been revolutionary, but, like most revolutions, the result was not what we expected it to be.

*Analysis of data*

The first applications of computers to forest research were to the analysis of experimental and survey data. This emphasis was inevitable, because, by the time that the computers were becoming available the need for analysis had already overstretched the simple computing devices that were available. Increasingly complex experimental designs had become necessary for the control of unwanted variation in experimental sites and material. Survey design had developed from the very simple systematic transects to the multi-stage and cluster-sampling designs required to improve the efficiency of sampling heterogeneous populations. Multivariate techniques of analysis were needed to investigate the influence of many factors on populations which could themselves only be described adequately by several variables. The computer, with the wide variety of algorithms that were developed quickly to enable such computation to be done rapidly and efficiently, and with the access to machine-readable data that resulted from the ever widening use of the available facilities, opened up new possibilities for analysis and interpretation of the many complex data sets which had so far defied analysis. Many methods that had been suggested for data analysis were found to be less useful than theory had suggested, and therefore new techniques were rapidly developed. Good accounts of the enormous surge of new ideas which followed this phase of computer application are given by Cooper (1969), Jeffers (1963, 1968), Nelder (1984) and Yates (1962).

Early approaches to data analysis tended to assume that the ideal would be a single pass of data through a previously selected program, and that the results could then be interpreted and written up from the output from this program. Ideally, the computer itself would produce most of the information necessary for the presentation of the results. We know now that efficient data analysis frequently requires much more than a single pass through the computer. Usually, it will be necessary to select data from a larger data-base, and then to explore these data in order to determine the main trends, including the presence of heterogeneity and heteroscedacity, the need for transformations to achieve approximate normality of the distribution of the data, and the presence of any non-linear relationships. Only then will it usually be possible to select an appropriate form of analysis, leading to the interpretation. Even after the analysis is complete, it may still be desirable to prepare a synthesis which will lead to some kind of model to be included in the final presentation of the results. The whole process of data analysis is therefore very much more complex than is commonly supposed, and requires an integration of statistical and computing skills with a knowledge of the experimental techniques.

*Forest mensuration*

There had always been a strong computational interest in forest mensuration, especially in the production of volume and yield tables, and in the calculation of yield increments and the harvestable proportions of the crop. Because of the difficulties of manual calculation, much of the practice of forest mensuration had been adapted to graphical and semi-quantitative techniques, and it might have been expected that these would quickly be replaced by the more direct computational methods made possible by the newly available computers. In fact, there was distinct lag between the computer becoming available and its enthusiastic use in forest mensuration. The graphical methods of traditional forest mensuration were too well-established, and it was not until some of the younger scientists who were trained in the use of computers moved into this field of application that the construction of yield and volume tables was rethought (Christie, 1972; Schopfer, 1978), although the fundamental basis for that change had existed for some time.

*Simulation and modelling*

A less expected application of computers was in the field of simulation. The concept of 'models' as the formal expression of relationships describing some theoretical or practical situation had been at the basis of much of the statistical and applied mathematics for some time (Maynard Smith, 1974), but computational solutions began to augment and replace the purely analytical solutions of conventional applied mathematics. As confidence was gained in the iterative solutions of difference equations, the scope and content of the models were expanded. Mathematical simulations of biological and ecological systems—including those of forestry—rapidly began to take an increasingly important role in forest research (Jeffers, 1982).

Simulation now forms a medium of communication between the experimental scientist and the resource manager or administrator. The model provides information for the data-bank used by the manager in his management of the system. The same model is used by the research scientist to test hypotheses before designing formal experiments in the field. This combination of simulation and modelling with the more traditional forms of field experiments is one of the unexpected consequences of using computers in forest and ecological research (Dent and Blackie, 1979).

Modelling and simulation, however, were not found to be successful when used in isolation. The ambitious research programmes of the International Biological Programme (IBP), for example, demonstrated all too clearly the need to embed simulation in a broader framework of thought and investigation. This broader framework was found in systems analysis, the orderly and logical organisation of data and information into models, followed by the rigorous testing and exploration of these models necessary for their validation and improvement (Jeffers, 1978).

*Graphics*

The early computers had only very limited facilities for the output of results. Indeed, in some of the earlier machines, users were actively discouraged from printing more than a single column of numbers down the left-hand side of the page, since the output from the machine was so slow. Gradually, however, the ability to produce an intelligible output increased until it was possible to print integrated tables and texts, together with simple graphs. The next stage in the development of computer hardware was towards the use of a CRT display, selected parts of which could be printed. Especially when combined with micro-processors, such displays provide a rapid and convenient method of presenting the results of calculations, while enabling the important and durable parts of the information to be selected for retention. Gradual improvements in the resolution of the displays has enabled us to develop facilities for the production of diagrams, figures, and drawings of great complexity, as well as in colour. Such graphical representations, combined with text and tables, have revolutionised our ability to present information. With the ability to retrieve and display information from magnetic media very quickly, or to perform new computations or logical procedures on the information, the way is now open virtually to dispense with paper as a medium of communication, especially as micro-processors are already easily portable, and are likely to become even more so in the future. Few of us, however, have yet reached this stage of confidence in what, to many, is still regarded as a new technology.

## THE FUTURE

What is likely to be the further impact of computers on forest research? To judge from the past few decades, it is not easy to predict precisely what this impact may be, but we may confidently expect the continuation of some existing trends. If the Japanese succeed in their ambitious Fifth Generation programme for micro-computers, the power of the machines will be increased by several orders of magnitude within five years, and the cost will be reduced by at least one order of magnitude in the same period. The majority of the world's computers will be micro-computers, but these will be increasingly more powerful and versatile. Many will undoubtedly be single-purpose machines, designed to be attached as ancilliaries to complex pieces of equipment to control and monitor the operation of the equipment, but the general purpose machine will probably continue to dominate the market.

At the moment, word processing and other office procedures tend to be separated from the more scientific uses of computers, but word processors are essentially computers adapted for a special purpose. Simple packages for word processing already exist for many personal micro-computers, and there seems little point in maintaining a separation between the functions of word processing and numerical computing or graphics. The future systems will be able to handle all three functions with equal ease.

Similarly, despite the present difficulties in communication between machines, there is now so much demand for easy communication between computers and word processors of all kinds that this will have become a standard requirement within a few years. It will then be possible to pass information of all kinds (including text, data-base and algorithms) from person to person over short or long distances. Given the ability to recall information easily, we may then see the gradual reduction in our present preoccupation with printed material. Especially because it is possible, with the aid of a computer, to interact with information which is already stored, and therefore to augment and modify it, a more flexible approach to learning and communication of research results can be expected to result. Such a development will be particularly influential as expert systems are built, as discussed below.

During the relatively short history of computers, there has been much discussion of the alternative advantages and disadvantages of the many programming languages that have been constructed for the purpose of writing the necessary algorithms to control the operation of computers. Today, one of the most popular languages for use on small, personal computers is BASIC, although computer scientists almost universally regard this language as inadequate. In the future, it seems likely that none of these arguments will be important. It actually matters very little which language is used, although some languages are more suitable for some purposes than others. It will not be long before the computers will themselves be programmed to translate programs from one language to another, thus leaving the maximum of choice to their human operators, who necessarily differ in skill and in their preference of language.

Whatever may be the fate of individual languages, there is little doubt that the algorithms will remain the most important product of the computer. The ability to express algorithms in a variety of different ways, including graphically, will almost inevitably increase, because understanding of the algorithm may frequently be more important than actually executing the statements. Already, distinctions are beginning to be made between descriptive and imperative computer languages, and languages like PROLOG (Clark and McCabe, 1984) have been developed to aid the programming of logical relationships.

One of the many imaginative uses of computers is that of exploring the mental models that individuals have of defined facets of the everyday world. Shaw (1980) has developed a series of computer programs which interrogate an individual, or group of individuals, so as to help them define constructs that they use to distinguish between sets of elements. The techniques embodied in these programs have proved to be exceptionally useful in improving communication between individuals, or between groups. Essentially, they avoid the misunderstandings that arise from the use of the same constructs to mean different things, or, alternatively, different constructs to mean the same thing. The use of computers as an adjunct to the human mind, and as an interrogator, raises some fascinating questions about the relationship of computing with intelligence, and also opens up some new avenues of research in forestry and many other practical fields.

However, perhaps the most exciting possibility for the future of computers in forest research, as in many other fields, is in the so-called expert systems. An expert system can be defined (Naylor, 1983) as:

> the embodiment within a computer of a knowledge-based component from an expert skill in such a form that the system can offer intelligent advice or take an intelligent decision about a processing function. A desirable additional characteristic, which many consider as fundamental, is the capability of the system, on demand, to justify its own line of reasoning in a manner directly intelligible to the enquirer.

A special style of rule-based programming is therefore required to attain these characteristics. Work on expert systems related to forest research is only just beginning, but, if the early promise of these systems is fulfilled, they will represent by far the most important contribution of the computer to forest research. While simulation and modelling have been added to data analysis in the improvement of research methods made possible by today's computers, a principal difficulty with conventional models is that they quickly become unintelligible to the forest managers and administrators who need to use them. Expert systems, in contrast, offer the possibility of providing a practical means of communication between research scientists and forest managers. They will also release human experts with scarce skills from providing day-to-day guidance, thus enabling them to improve and extend their expertise; these extra skills can then themselves be embodied as expert systems.

## POSTSCRIPT

Some years ago, my grandfather said to me how grateful he was to have lived through the period of history during which man had conquered the problems of travel. During his life-time, he had seen the development of the motor vehicle, the aeroplane, and even the beginnings of space travel. He had felt it a privilege to have lived through so exciting and so challenging a period. I, in my turn, feel much the same, having witnessed, and played some small part in, the development of computing in forestry. This development has removed the constraints of computation from man's use of his intellect and, at the same time, given him new tools with which to extend his control and understanding of our natural resources. It has been an enthralling and mind-stretching experience to have lived through such a rapid technological and theoretical change. However, the most exciting prospect of all is that we are only just beginning on an intellectual revolution whose consequences for change will far outstrip those of the industrial and transport revolutions.

REFERENCES

Christie, J.M. (1972) The characterisation of the relationships between basic crop parameters in yield table construction. In: *Proc. 3rd Conf. Advisory Group of Forest Statisticians, Sect. 25.* Paris: IUFRO.

Clark, K.L. and McCabe, F.G. (1984) *Micro-PROLOG: programming in logic.* Hemel Hempstead: Prentice-Hall International.

Cooper, B.E. (1969) Statistical computing—past, present and the future. *Statistician* **19**, 125-141.

Dent, J.B. and Blackie, M.J. (1979) *Systems simulation in agriculture.* London: Applied Sceince.

Hummel, F.C. (1955) The volume-basal area line: a study in forest mensuration. *Bull. For Commn., Lond.* **24.** London: HMSO.

Jeffers, J.N.R. (1963) The electronic digital computer in forest research and management. *Forestry Commission Report on Forest Research, 1962* 166-178.

Jeffers, J.N.R. (1968) Application of electronic digital computers to forest research and management. *Forestry Abstracts* **29**, 209-216.

Jeffers, J.N.R. (1978) *An introduction to systems analysis: with ecological applications.* London: Academic Press.

Jeffers, J.N.R. (1982) *Modelling.* (Outline studies in ecology.) London: Chapman and Hall.

Maynard Smith, J.M. (1974) *Models in ecology.* Cambridge: Cambridge University Press,

Naylor, C.M. (1983) *Build your own expert system.* Wilmslow: Sigma Technical Press.

Nelder, J.A. (1984) Present position and potential developments: some personal views. Statistical computing. *Journal of the Royal Statistical Society A* **147** 151-160.

Schopfer, W. (1978) Forest biometry and computer science in practice, illustrated by current projects of a forestry research institute. *Forstwissenschaftliches Centralblatt* **95**, 226-243.

Shaw, M.L. (1980) *On becoming a personal scientist.* London and New York: Academic Press.

Yates, F. (1962) Computers in research—promise and performance. *Computer Journal* **4**, 273-279.

# DATA ANALYSIS USING A MICRO-COMPUTER

D.K. LINDLEY

*Institute of Terrestrial Ecology, Merlewood Research Station, Grange-over-Sands, Cumbria LA11 6JU.*

SUMMARY

Easy access to micro-computers today means that complex calculations that took three weeks to calculate on an electric calculator can now be computed in a few seconds. Although the heart of data analysis rests in series of instructions or procedural steps, called algorithms, the user must be concerned with the efficient methods of data input and storage, with displaying the results in an interesting and informative manner and with production of printed results. Efficient data analysis on a micro-computer must consider an optimum method of linking these several stages together. Example data sets will be used to illustrate how one widely used statistical method, multiple regression, can be linked with a purpose-built data input and editing program to enable the user to carry out this method of analysis on the BBC micro-computer.

## INTRODUCTION

Many students and recently graduated foresters will be used to working with computer systems. By now, all universities and polytechnics have a range of micro-computers, in addition to their large central main-frames; both types are used for teaching, research and administrative purposes. Those foresters working for local authorities, research organisations and large forestry companies will generally have access to sets of programs or packages on centrally located computers.

By contrast, the forester on the small forestry estate is most unlikely to have access to such computing facilities. However, even these foresters need not be deprived of modern methods for data analysis, since they are available on an inexpensive micro-computer.

Once an individual or a small organisation has identified the need for some form of data analysis facilities, a difficult decision about what hardware and software to purchase has to be made. The key to any choice should be flexibility. You may be interested in data analysis techniques to store and summarise your data and to enable you to look for relationships within your data sets but you may also want to do word processing, accounting, forecasting and production of educational material as well as other topics. One system may be excellent at word processing but weak on accounting so you should evaluate your choice very carefully.

## STATISTICAL SOFTWARE FOR MICROS

A very recent and useful review of statistical computing is contained in the Professional Statistician, Volume 3, No. 7. Gilchrist (1984) mentions that there are several reports on statistical software for micro-computers readily available; for example, Neffendorf (1983) provides a list of about 50 micro packages; Cable & Rowe (1983) have also listed about 50 packages. These two listings when combined together give nearly 70 packages.

Gilchrist also comments that statistical software may not be available for many machines at the lower end of the market. Many of the British-based 'home computers' have no substantial statistics software available. This situation is likely to change in the near future with some software houses turning away from games to more 'useful' and serious applications.

When examining the existing statistical software it appears that most of the hardware and packages together start at over £2 000. Most of these systems are designed for the business or scientific market.

## AN EXAMPLE OF DATA ANALYSIS ON THE BBC MICRO

### *The BBC micro-computer*

This micro-computer can be used for data analysis. It is a flexible machine which can also be expanded to a larger business system at a later stage; it is supported by a large user group; it is well known in schools, universities and colleges and there is a rapidly-expanding range of software becoming available for it. In the autumn catalogue from Acornsoft, 13 business packages were listed. One or two statistical packages are now beginning to appear on the market.

No matter what micro system you use for serious business and scientific purposes, there will generally be what is said to be a 'minimum configuration'. In other words, it is the minimum amount of hardware that is required to produce a useful working system. In the current example, it will be a 32K Model B machine, a monochrome monitor, a printer and a floppy disk unit which can be put together for under £1 500. A disk system is required because of its high speed of operation and because it allows one program to call another allowing the user to deal with much more complex operations, (see Sinclair, 1984, for a good presentation of disk systems on the BBC micro). A printer is essential so that data and the results can be printed out.

### *A data analysis program for the BBC micro-computer*

My colleague Dr P. Bacon has developed a general purpose data input and editing program written in BASIC for the BBC micro-computer and called DATFILE. The program is menu driven, allowing the user to jump into an appropriate part of the program depending upon the response at specific stages of the program. In general terms, the user must be able to enter data into the computer from the keyboard; mistakes are bound to be made and therefore some means of correcting the mistakes must be available; a 'hard copy' or print-out will be required so that the original data set can be checked against the set stored in the computer. Finally, the ability to produce a data summary has been found to be useful in finding further mistakes.

With the BBC micros, as with most other makes, any programs or data that are residing in the memory are lost, when the system is turned off. Floppy disks enable the user to store data and programs as named files. The DATFILE program has the ability to write data to a file and to retrieve data from a file stored on the disk. Other useful options include the handling of missing data, allowing for early and unplanned ending of data input and the use of a one-line title to describe the data set stored in the file.

*A typical data analysis problem*

The example that will be considered is that of a hypothetical private forest estate where the forest manager has been collecting data about the costs of forest operations for a number of years (see Lindley & Philip, 1968 for specific examples). The forest manager will always be involved with financial planning and there is often no shortage of information available to help him make his decisions. However, if the cost of a forest operation is influenced by more than one factor then he could find the statistical technique, multiple regression, useful to help him obtain more precise estimates of cost.

A useful introduction to the statistical theory of the multiple regression technique is provided by Snedecor & Cochran (1980). Essentially, it is an extension of the straight line equation $Y = a + bX$. The multiple regression program will be used to calculate the constant and coefficients for the following model:

$$Y = a + b_1X_1 + b_2X_2 + b_3X_3 + b_4X_4 + b_5X_5 + b_6X_6$$

where $X_1 \ldots X_6$ are the six independent variables. Some information about suitable programs in BASIC for the multiple regression technique is provided by Cooke *et al.* (1982).

In order to use the technique of multiple regression, costs and the related variables have to be collected together in a form suitable for input into the multiple regression program. However, a set of data on costs may be scattered throughout a number of compartment records. Even if the estate is advanced in its office operations and has its compartment records already on a computer file, the relevant data items will still need to be extracted from the main file and stored in a separate computer file. This example assumes that the data are manually extracted and recorded onto a data form. The record for each compartment will be stored as one row; the variables will be the columns.

In this example, the labour cost for planting each compartment, expressed in pounds per hectare, is thought to be influenced by the following factors:

Area planted (in hectares);
Total number of plants used (in hundreds);
Factor for spacing;
Coded value for the species planted;
Coded value for the age of plants;
Coded value for type of planting.

James (1966) suggests that the wider spacing, the less plants are needed, and therefore the cost of planting is reduced. The factor for spacing reflects this idea and represents the square of the average distance between the plants:

$$\text{Plant spacing factor} = (10\,000/\text{Number of plants per hectare}) \times 100$$

Early work by Aberdeen University on their Economic Survey of Private Forestry in Scotland indicated that type of species and method of planting could have some influence on the cost of planting. The problem arises as to how to quantify species type and the method of planting. If different methods such as plough, direct-notch, and turf planting are thought to take different amounts of time in planting they can then be ranked in order and given specific values to suggest that one method takes longer than another. We have 3 different weights for method of planting in our example 2, 5 and 7. The same approach has been applied to quantify the effects of different species.

```
You may: Input data from Keyboard   = I
         Edit the data file         = E
         Print the data out         = P
         Get summary statistics     = S

         Write the data to a file   = W
         Retrieve data from file    = F
         Append data to OLD file    = A
         Terminate this program     = T

(reply as CAPITAL letter; RETURN)
What do you want to do?
I, E, P, S, W, F, A, T ?I
```

Figure 1. The menu for program DATFILE

If no information exists about the possible weights that can be attached to the individual items of a qualitative variable, then the use of dummy variables should be employed. For example, independent variable 6, the type of planting, could be expanded into two dummy variables, say X6 and X7, where they are used for identification of the type of planting as follows:

| X6 | X7 | |
|---|---|---|
| 1 | 0 | = plough |
| 0 | 1 | = direct-notch |
| 0 | 0 | = turf |

See Draper and Smith (1981) for a useful explanation of the more complicated models that can be used in developing multiple regression equations.

*Use of the program for the analysis*

These data are extracted from the source documents and recorded on a collation sheet.

Before the DATFILE program can be activated, it will need to be copied from the floppy disk into the computer memory and then RUN. The menu will then appear on the screen as shown in Figure 1. The program provides a number of prompts; it needs to know the maximum number of rows (compartments) and columns (variables) so that sufficient space can be reserved in memory. In this example, data from 70 compartments are available for the dependent variable, planting costs, plus the six independent variables. Prompts are then provided to guide the user in entering the data one row at a time, see Figure 2. Once the data have all been entered, a warning message is given and the menu re-appears. It is valuable at this stage to print the data out and also to produce a summary table, see Figure 3 and 4. If errors are found they can be corrected by using the EDIT option. The final stage of data preparation is to transfer the data existing in memory to a named file which will permanently reside on the disk. A certain amount of care is needed at this stage to make certain that the file names are useful and understandable.

It is good practice to store the analytical programs separately from the data in order to guard against giving a data file the same name as a program. If this unfortunate event happens the program is overwritten. Therefore, it has to be retrieved from its own floppy disk.

Because the multiple regression program is quite long (247 lines of BASIC code), it has been written in 3 sections; 2 and 3 being called automatically once the first

```
Maximum number of Rows          ?70
Maximum number of Columns       ?7
value that signifies missing data  -999    ?-999

value to allow premature ending of data input    -999999    ?-999999
a title line to describe this file
? COST OF PLANTING

Row 1
Col.
          1 value ?44.18
          2 value ?4.45
          3 value ?140
          4 value ?336
          5 value ?2
          6 value ?7
          7 value ?2

Row 2
Col.
          1 value ?35.77
          2 value ?3.64
          3 value ?100
          4 value ?397
          5 value ?2
          6 value ?7
          7 value ?2
```

Figure 2. Entering the first 2 records

Cost of planting

| Row No. | Cost (£) | Area (Ha.) | No. plants (00) | Spacing | Species Code | Age Code | Planting Code |
|---|---|---|---|---|---|---|---|
| 1 | 44.18 | 4.45 | 140 | 336 | 2 | 7 | 2 |
| 2 | 35.77 | 3.64 | 100 | 397 | 2 | 7 | 2 |
| 3 | 26.30 | 8.10 | 194 | 449 | 2 | 2 | 2 |
| 4 | 69.43 | 16.19 | 513 | 336 | 5 | 2 | 5 |
| 5 | 32.61 | 4.86 | 115 | 449 | 7 | 2 | 2 |
| 6 | 26.30 | 0.81 | 20 | 436 | 2 | 7 | 2 |
| 7 | 44.18 | 9.31 | 376 | 260 | 5 | 2 | 2 |
| 8 | 45.24 | 7.29 | 310 | 250 | 5 | 2 | 2 |
| 9 | 38.92 | 4.86 | 171 | 303 | 2 | 2 | 2 |
| . | | | | | | | |
| . | | | | | | | |
| . | | | | | | | |
| 70 | 67.28 | 0.81 | 19 | 348 | 7 | 2 | 5 |

Figure 3. Printout of data set using option P

segment is finished. Section 1 calculates the sums of squares and products, 2 works out the regression statistics whilst the last part plots the residuals.

The multiple regression program needs to know the name of file where the data are stored on the disk: in this example the file is called PLANT (Figure 5). Once the calculations are completed, the results are printed out with the data file name, the

Calculating Stats

SUMMARY TABLE for 70 data lines

| Column 1 | |
|---|---|
| Number of items | 70 |
| Mean value | 44.543 |
| Standard deviation | 13.017 |
| Minimum value is | 26.30 |
| Maximum value is | 85.44 |

| Column 2 | |
|---|---|
| Number of items | 70 |
| Mean value | 10.723 |
| Standard deviation | 26.402 |
| Minimum value is | 0.40 |
| Maximum value is | 179.35 |

Figure 4. Statistical summary for the data set, variables 1 and 2 only

Multiple regression

Data file: plant
6 Independent variables and 70 sets

| COEFF | Estimate | Standard error |
|---|---|---|
| A | 67.9672 | |
| B(1) | −1.2983 | 0.8198 |
| B(2) | 0.0426 | 0.0264 |
| B(3) | −0.0991 | 0.0188 |
| B(4) | −0.4811 | 0.5789 |
| B(5) | −0.6000 | 0.7831 |
| B(6) | 4.5601 | 0.7954 |
| R Squared = 0.5931 | | |

Figure 5. Multiple regression results

number of independent variables and the number of sets acting as a heading, as in Figure 5. The results indicate that 59% of the variability in the labour cost of planting can be 'explained' by these 6 variables. The predictive equation is:

$$Y = 67.97 - 1.30X_1 + 0.04\,X_2 - 0.10X_3 - 0.48X_4 - 0.60X_5 + 4.56X_6$$

Further study of the ratio between the estimate of the coefficient and its standard error indicates that only 2 of the 6 variables make a significant contribution, namely the spacing factor, $X_3$, and the type of planting $X_6$. Re-calculating the equation using only these two variables provides the simpler predictive equation:

$$Y = 71.72 - 0.12X_3 + 4.39X_6$$ (57% of the variability 'explained')

If there is a compartment where the spacing factor is to be 290 and the type of planting is code 5, then the predicted labour cost of planting per hectare is calculated as:

$$71.72 - (0.12 \times 290) + (4.39 \times 5)$$
$$= £58.87$$

If instead of using weights 2 and 5 to represent two different types of planting in the above example, a dummy variable is used then the following equation is obtained:

$$Y = 80.49 - 0.12X_3 + 13.16\,X_6$$

It has the same predictive power as the previous equation although the regression constant and coefficients are changed.

## DISCUSSION

Although multiple regression is a very useful and powerful statistical technique, great care should be exercised in applying to it any set of data. Much thought should be given to possible intercorrelations between the 'so-called' independent variables, and more complex models using transformations (e.g. logarithmic) should be explored before deciding that the definitive predictive equation has been obtained. It is also a useful exercise to test the equation out on a second, independent set of data.

People often underestimate the length of time that it takes to analyse some data because of the time required for the preparation and data input stages. This misjudgement often takes place in scientific research where, although the data sets may be relatively small by commercial standards, they are often spread over a number of different forms or notebooks and generally require some form of further manipulation before being ready for the analytical stage.

The two programs presented in this paper do not constitute a full analytical 'package'. However, their use does illustrate many of the features that are required of a package. Data have to be entered into the computer by some means, a correct algorithm has to be found for a specific analytical technique, long term data storage should be available and finally some permanent printed record should be kept both of the original data and of the results.

There are many potential uses for data summary and analysis programs on the small forest estate. They can be used to highlight anomalies as well as patterns in your data. I suggest that it is a useful investment of time to find out about them for your own particular micro-computer, or even investing in a micro-computer with a suitable range of packages, one of which should be for data analysis.

## ACKNOWLEDGEMENTS

Thanks are due to my colleagues, Dr P.J. Bacon and Mr P.J.A. Howard for developing the programs mentioned in this paper.

## REFERENCES

Acornsoft, (1984) *Spectacular programs for the BBC microcomputer from Acornsoft.* Cambridge: Acornsoft.

Cable, S. and Rowe, B. (1983) *Software for statistical and survey analysis.* London: Study Group on Computers in Survey Analysis.

Cooke, D.; Craven, A.H. and Clarke, G.M. (1982) *Basic Statistical Computing.* London: Arnold.

Draper, N.R. and Smith, H. (1981) *Applied Regression Analysis.* 2nd Edition. New York: Wiley.

Gilchrist, R. (1984) The Polytechnics' Micro-Evaluation Project. *The Professional Statistician, 033, 7,* 17-19.

James, N.D.G. (1966) *The Forester's Companion.* Oxford: Blackwell.

Lindley, D.K. and Philip, M.S. (1968) Economic surveys of forest enterprises. Presented at the 9th British Commonwealth Forestry Conference.

Neffendorf, H. (1983) Statistical packages for microcomputers: An update. *The Professional Statistician,* **2, 6,** 8-11.

Sinclair, I. (1984)*Disk Systems for the BBC Micro*. London: Granada.
Snedecor, G.W. and Cochran, W.G. (1980) *Statistical Methods*. 7th Edition. Ames: Iowa State University Press.

# BIO-FUEL UTILISATION: A SIMPLE EXAMPLE OF A COMPUTER SIMULATION MODEL

P.J. BACON

*Institute of Terrestrial Ecology, Merlewood Research Station, Grange-over-Sands, Cumbria LA11 6JU.*

SUMMARY

Models play a vital role in everyday human communication and understanding; ideas, words, maps, diagrams and working scale models are all simplified representations of real objects that can, very broadly, be termed 'models'. The limitations and benefits of modelling are briefly described, and a series of diagrams used to explain how a real problem, (the provision of fuel wood in a developing country), can be simplified into a form in which the precision of mathematics and the calculating speed of a micro-computer can be combined to give useful insight into complicated processes. Example data are used to indicate how the model's predictions in one set of circumstances can be used to suggest management options: the controlling data and formulae can then be modified to reflect these 'managed' circumstances and the model re-used to predict the consequences of that 'management' and hence allow its evaluation.

## INTRODUCTION

Models can be usefully defined as 'a simplified representation of the real world' and as such can contain varying degrees of GENERALITY, REALISM and PRECISION in relation to the entities, attributes and processes of the real world which they represent. However, the basic act of SIMPLIFICATION requires that some degree of one or all these aspects is sacrificed in a model. For example, very high degrees of Realism and Precision will prevent a model being General while a highly Realistic and General model could not be extremely Precise. The art of modelling requires that the objectives of the model be clearly defined so that the assumptions and formulae that represents the simplification process can be made adequate and appropriate for those purposes (Jeffers, 1978; Holling, 1978). Users of models (including economic forecasts) should realise that the model is only as good as the assumptions and data which underlie it, and they should have an adequate understanding of what these are: the fact that a model is implemented by a computer does not ensure it is useful, appropriate or accurate.

## THE EXAMPLE: PRODUCTION OF FUEL-WOOD

### *Background and Objectives*

In developing countries, fuel wood is vitally important for cooking and may provide for over 90 percent of a country's total energy usage. Population growth, a sluggish economy and lack of natural resources often result in a vicious downward spiral:— over-harvesting, tree death, soil erosion, desertification, reduced area of forest, reduced yield of fuel wood, . . .

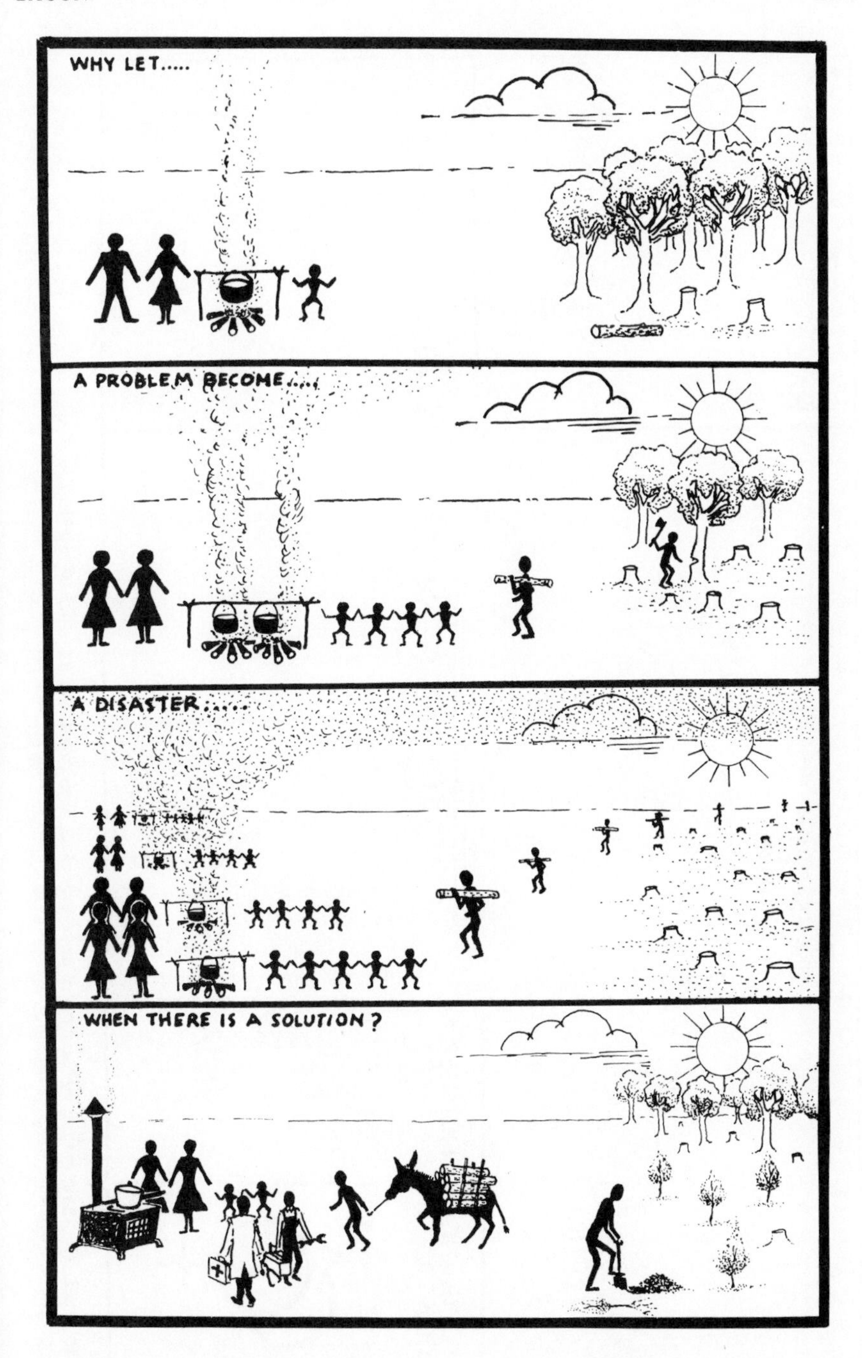


FIGURE 1. Diagrams of a progressively worsening unmanaged situation (a-c) and a possible managed solution (d).

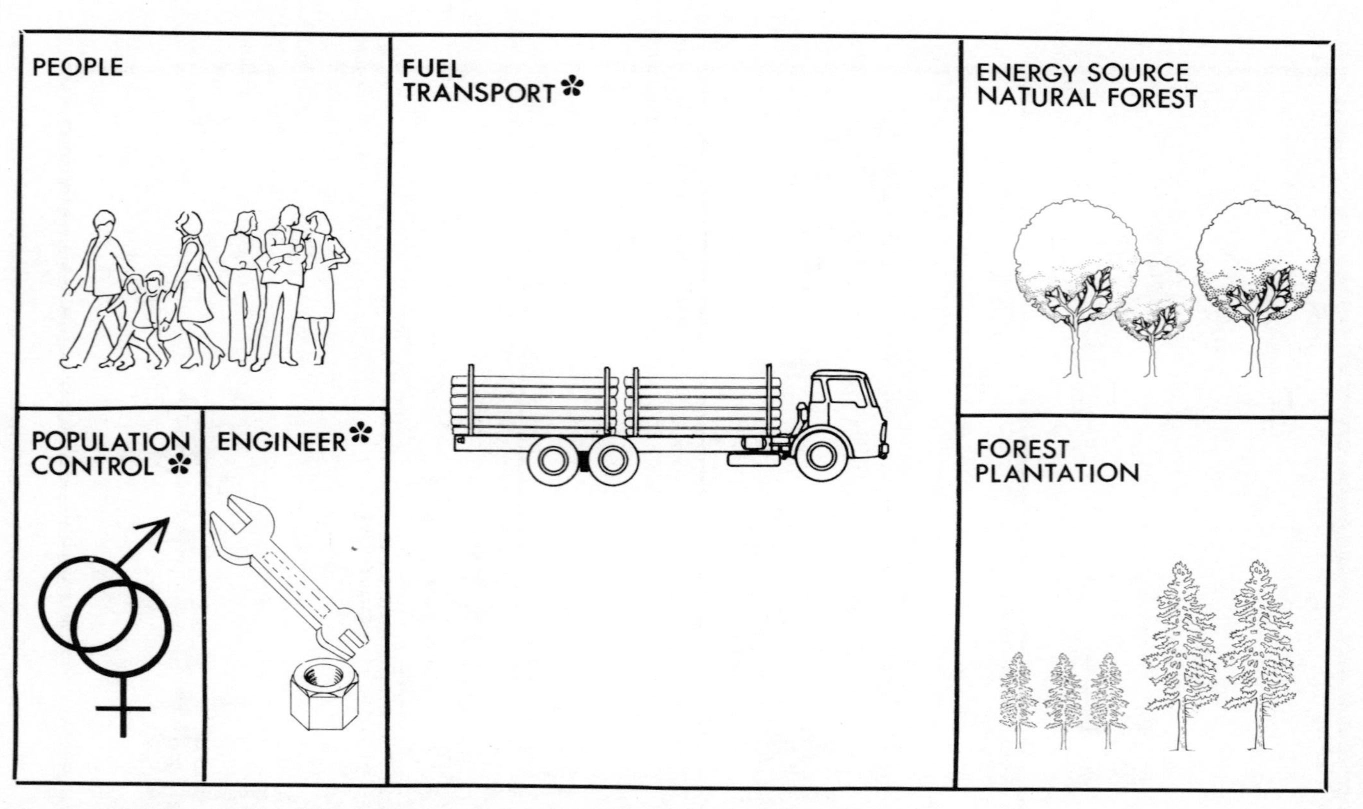


FIGURE 2. The main entities of the system to be modelled, with those that are only included implicitly shown with an asterisk. All other compments are actually modelled.

The model described here was developed as a teaching aid to illustrate the (approximate) biological consequence of different management policies aimed at alleviating the fuel wood shortage. These objectives require generality and can be met with low degrees of realism and precision. The essence of the problem is illustrated in the diagram of Figure 1. For present purposes the complexities of (a) transport of fuel and (b) the economics of the costs and benefits of the management options are omitted (we are presently only interested in the biological consequences), but these could be added later if needed. Accordingly, the main entries of the whole system, as shown in Figure 2, can be further simplified to those highlighted in the figure. Details of the flows of energy are indicted in Figure 3.

Adjustable energy losses are: (i) harvest—small branches usually wasted could be pelletised; (ii) transport—improve transport efficiency or grow trees nearer sources of use; (iii) conversion to charcoal—earth kilns only about 15 percent efficient, modern kilns 30-40 percent efficient, or burn wood; (iv) primitive stoves only 10-15 percent efficient, slightly more expensive ones can be 30-40 percent efficient. Forest or plantation management, by irrigation, fertiliser and new species, could also improve initial yields.

Our aim is to use the model to investigate how, within the biological constraints the model imposes, the highest proportion of available energy from the sun can be utilised, and to assess whether or not this will meet the needs of a country with a changing population size.

*Model structure*

The entities and processes of the model are summarised in Table 1. Table 2a lists the 'management options' we wish to investigate and Table 2b shows how these options correspond to the various 'input' sections where parameter values and formulae are supplied to the model.

*Model implementation*

The model was implemented on a BBC micro-computer in BASIC. This choice of computer and language resulted from three considerations:

i) they were available at both Institutes involved in the project (the Institute of Terrestrial Ecology and the Institute of Environmental Studies (IES), University of Khartoum, Sudan);
ii) the EVAL (formula) statement in BBC BASIC allows both data values and formulae to be changed by the user (not just data as in many models);
iii) BBC BASIC is a powerful interactive language with good colour graphics and allowing long meaningful, variable names.

The structure of the program implementing the model is shown as a logical flow diagram in Figure 4. Within the model the calculations are performed by 'formulae' which, in BBC and other good versions of BASIC, can be made easy to understand. For example, the total forest growth in a year is calculated by the statement

```
F_GROW_THIS_YR = F_LAND * F_GROW
```

| where | F_ | is a prefix signifying forest |
|---|---|---|
| | = | means 'is set equal to' |
| | * | means 'multiply by' |
| | F_LAND | is a variable holding the area of forest (Ha) |
| | F_GROW | is a variable holding the nett yield (tonnes/Ha) |

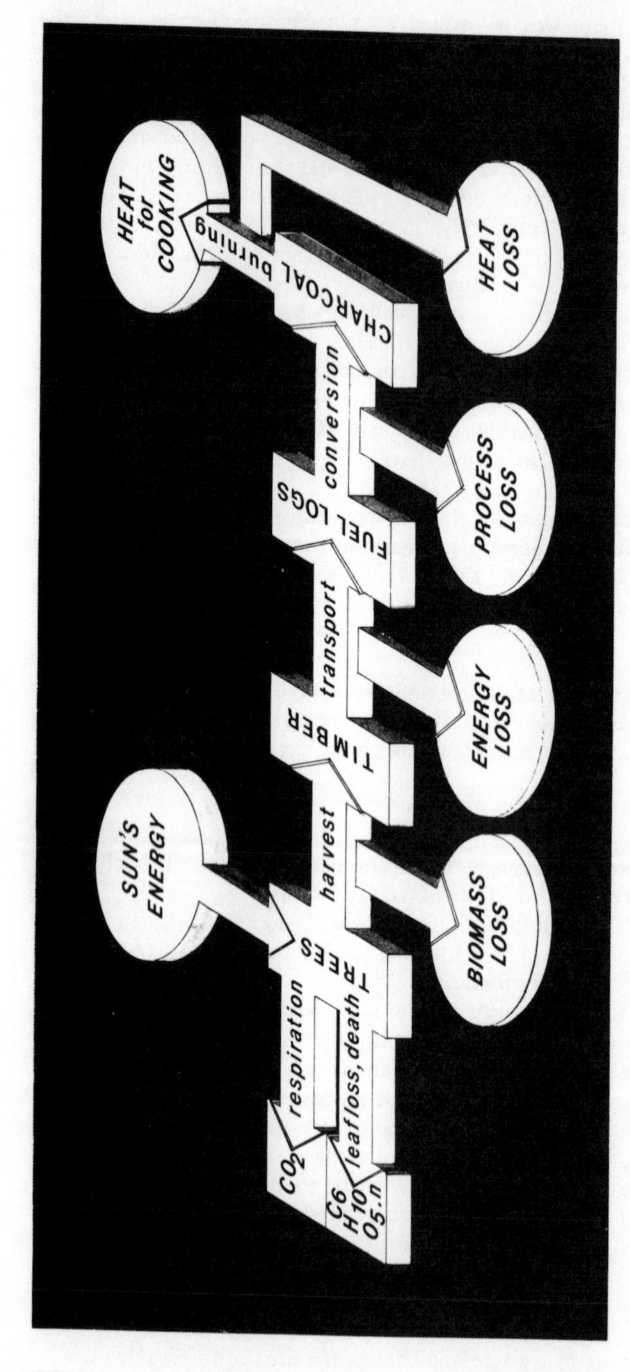


FIGURE 3. A formal diagram of the energy transfer from the source, (the sun) to the end use (cooking) showing the stages and routes whereby utilisable energy is 'lost'.

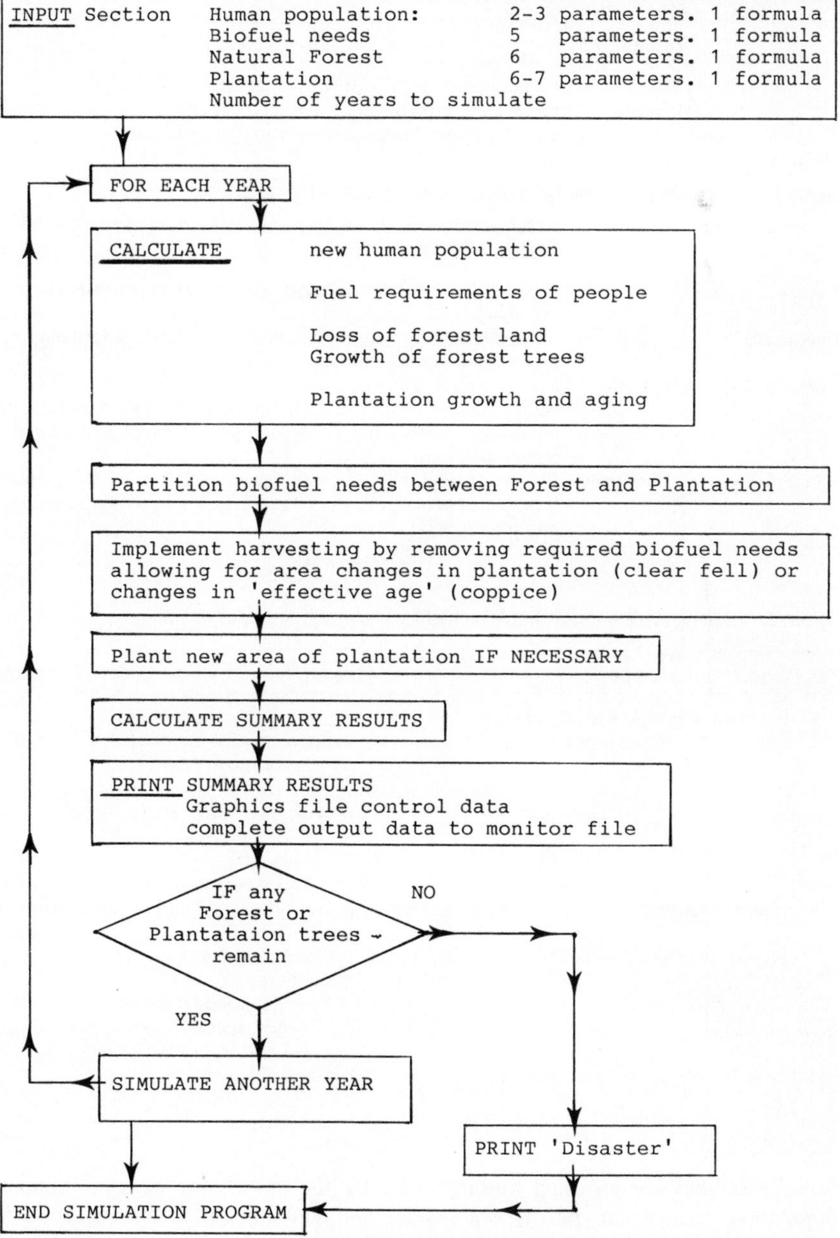


FIGURE 4. A 'flow diagram' indicating the sequences of calculations performed by the computer simulation program. Note that the calculations within the 'loop' arrows (from 'FOR EACH YEAR' to 'SIMULATE ANOTHER YEAR') is repeated a specified number of times.

Table 1a. Entities of the model with example attributes.

| | Entities | Attributes |
|---|---|---|
| 1 | People | Population size: growth rate. Fuel needs |
| 2 | Natural forest | Land-area; growth-rate; carrying-capacity |
| 3 | Plantations | Species; growth-rates; areas; management |

Table 1b. Processes in the model, indicating which entities these affect.

| Processes | Entities affected |
|---|---|
| Population growth | Population size affects population increase (via per capita births per year). |
| Fuel needs | Total human population size and their life style determine fuel needs to be met by harvesting biomass. |
| Forest growth dynamics | Forest growth dynamics depend on tree species composition, environmental factors, etc. Net growth is modified by human harvesting which will itself depend on fuel needs and whether plantations exist. |
| Plantation growth dynamics | Plantation dynamics depend on species, the environmental factors and management practice (including harvesting methods and intensity. |

Table 2a. Management options, implemented by changing data and formulae in the model.

| | ENERGY BUDGET OPTIONS | IMPROVEMENT POSSIBLE BY . . . |
|---|---|---|
| i | Number of people needing energy | control human population |
| ii | Energy requirements per person | improve efficiencies of energy use<br>—charcoal<br>—stoves |
| iii | Reduce demand on biofuels | provide alternative energy sources<br>—hydro-electric<br>—bio-gas<br>—petroleum |
| iv | Preserve natural forest as an energy source | prevent over-exploitation and maintain annual yields. |
| v | Provide plantaition as additional energy sources | provide additonal fuel-wood from plantations.<br>—increased land area<br>—new species<br>—improved growth rates |
| vi | Improve energy yield of plantations | method of harvesting<br>—clear fell<br>—coppice |

Some formulae are provided automatically by the model: for example human population growth can be (a) exponential and unbounded or (b) logistic and asymptotic, as illustrated by the relationships shown in Figures 5 (a and b respectively), and achieved by formulae in BASIC such as:

a) POP_SIZE = POP_SIZE + (POP_SIZE * BIRTH_RATE)

b) POP_SIZE = POP_SIZE + (POP_SIZE * BIRTH_RATE)*
(1–POP_SIZE/POP_LIMIT)

Table 2b. Simplified input dialogue between user and BIOEN1 simulation program.

Note: the characters in **bold type** were typed by the user.

| Parameters describing human population | |
|---|---|
| Number of people at start: | **100** |
| Per capita growth rate for people: | **0.02** |
| Human population growth function | |
| 1—exponential | |
| 2—logistic | |
| 3—other (user specified) | |
| Your choice: | **1** |

| Parameters determining bio-fuel requirements per person | |
|---|---|
| Efficiency of conversion: | **0.15** |
| Efficiency of users' stoves: | **0.30** |
| Energy content of biomass: | **18000** (joules per gram) |
| Energy needs per person: | **46000000** (joules per day) |
| Proportion of energy from biofuel: | **0.95** |

| Parameters describing the existing resource of natural forest | |
|---|---|
| Area of forest at start: | **1000** (hectares) |
| Forest land lost per year: | **20** (hectares per year) |
| Proportion of standing crop on lost land used for biofuels: | **0.45** (tonnes) |
| Biomass of forest at start: | **10000** (tonnes) |
| Maximum forest biomass: | **20** (tonnes per hectare) |
| Maximum forest growth rate: | **2** (tonnes per hectare per year) |

| Parameters describing tree planting and harvesting | |
|---|---|
| Maximum land available for planting: | **500** (hectares) |
| Area planted per year: | **20** (hectares) |
| Maximum plantation biomass: | **30** (tonnes per hectare) |
| Maximum forest growth rate: | **2.5** (tonnes per hectare per year) |
| Minimim age before harvest: | **4** (years) |
| Harvesting method: | |
| 1—clear-fell (kills trees) | |
| 2—coppice (allows regeneration) | |
| Your choice: | **2** |
| Time between successive harvests: | **2** (years) |

*Results from the Model*

The model automatically produces a table of summary information designed to allow the user to quickly evaluate how successful the strategy has been, and this information is here illustrated as Table 3. A basic goal of any strategy should be to ensure that demands (col. 3) become less than or equal to total growth (col. 4) (NB col. 8 = col. 4—col. 3 and hence should be, or become, positive for a successful strategy): demand greater than growth can only be tolerated as a short-term deficit. The results shown in Table 3 are for illustrative purposes only (see Bacon, Callaghan and Lindley for further details).

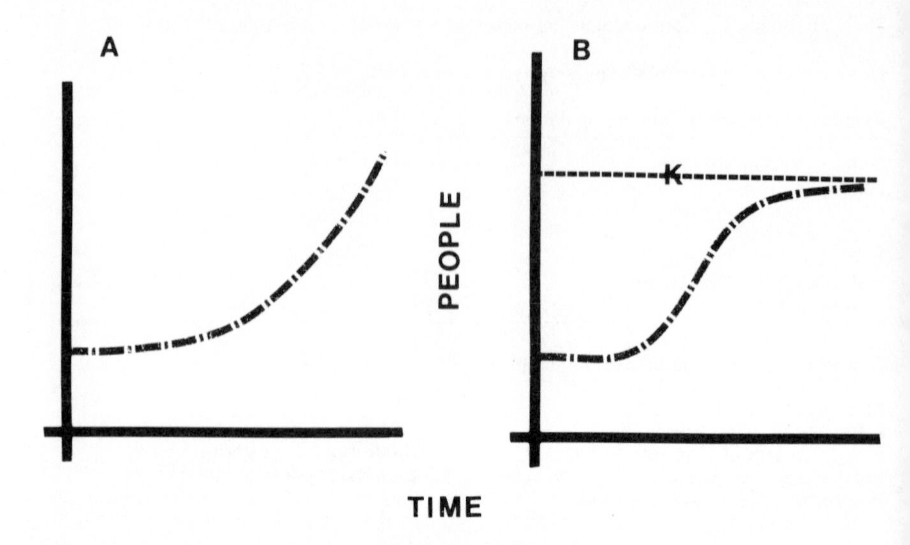


FIGURE 5. Illustration of the form of relationships that can be chosen to represent the increase of the human population with time during the simulation.
5.a Exponential, unbounded, increase
5.b Logistic increase to a specified level K, the population limit or 'carrying capacity'.

A more important point to emphasise is that once the computer has generated these results, other programs can be written to aid their presentation and interpretation. To facilitate such flexibility the program creates two output files of results, one to allow graphical presentation of the course of events that could be used to illustrate the main points to politicians and decision makers with little biological knowledge, and another to keep a complete record of the simulated changes (21 output variables for each simulated year). A graphics program uses the first file and 'cartoon' symbols to illustrate the results of the simulations in pictorial form varying with the years of the simulation. The 'monitor' file could be used to obtain more detailed tabulations of the results, to produce computer drawn graphs summarising those results or as input data to an 'economics assessment' program which could be written to assess the financial implications of the modelled strategy.

## DISCUSSION

The ability to use micro-computers to help present and explain the results of complicated scientific research is potentially most important as experience has shown that, if tackled in the right way, such models can provide a most fruitful means of promoting understanding and confidence between researchers and managers. The most important general lesson with regard to promoting such understanding is that managers and scientists should interact throughout the model design, development, testing, modifying and assessment stages (Holling, 1978; Jeffers, 1978). This close cooperation promotes mutual confidence, an understood and agreed terminology, and continually ensures that the model the researchers produce does what the managers require.

Table 3. Standard print-out to indicate the outcome of the model. The column headings explain what the output variables are (number of people, fuel biomass needed, etc) and the rows give the values of those variables for successive years of the simulation.

Note that for a successful long-term strategy column eight (excess of total growth over biomass needs) should be, or become, positive.

| | Human Popn. numbers | Fuel Biomass Needed tonnes | Total Biomass Growth tonnes | Forest Growth tonnes | Plantn. Growth tonnes | Total Biomass Now tonnes | Exces Growth-Needs tonnes | Excess Gross-Needs tonnes | % Land in use For. % | Plt % |
|---|---|---|---|---|---|---|---|---|---|---|
| YEAR | | | | | | | | | | |
| 1 | 102 | 2020 | 1960 | 1960 | 0 | 9830 | -60 | 9740 | 98 | 4 |
| 2 | 104 | 2060 | 1970 | 1920 | 50 | 9630 | -90 | 9540 | 96 | 8 |
| 3 | 106 | 2100 | 1980 | 1880 | 100 | 9390 | -120 | 9300 | 94 | 12 |
| 4 | 108 | 2150 | 1990 | 1840 | 150 | 9130 | -160 | 9040 | 92 | 16 |
| 5 | 110 | 2190 | 2000 | 1800 | 200 | 8840 | -190 | 8750 | 90 | 20 |
| 6 | 113 | 2230 | 2010 | 1760 | 250 | 8510 | -220 | 8420 | 88 | 24 |
| 7 | 115 | 2280 | 2020 | 1720 | 300 | 8150 | -260 | 8070 | 86 | 28 |
| 8 | 117 | 2320 | 2030 | 1680 | 350 | 7760 | -290 | 7680 | 84 | 32 |
| 9 | 120 | 2370 | 2040 | 1640 | 400 | 7330 | -330 | 7250 | 82 | 36 |
| 10 | 122 | 2420 | 2050 | 1600 | 450 | 6870 | -370 | 6800 | 80 | 40 |
| 11 | 124 | 2460 | 2060 | 1560 | 500 | 6380 | -400 | 6310 | 78 | 44 |
| 12 | 127 | 2510 | 2070 | 1520 | 550 | 5850 | -440 | 5780 | 76 | 48 |
| 13 | 129 | 2560 | 2080 | 1480 | 600 | 5290 | -480 | 5230 | 74 | 52 |
| 14 | 132 | 2620 | 2090 | 1440 | 650 | 4700 | -530 | 4640 | 72 | 56 |
| 15 | 135 | 2670 | 2100 | 1400 | 700 | 4060 | -570 | 4010 | 70 | 60 |
| 16 | 137 | 2720 | 2110 | 1360 | 750 | 3400 | -610 | 3350 | 68 | 64 |
| 17 | 140 | 2780 | 2120 | 1320 | 800 | 2700 | -660 | 2660 | 66 | 68 |
| 18 | 143 | 2830 | 2130 | 1280 | 850 | 1960 | -700 | 1930 | 64 | 72 |
| 19 | 146 | 2890 | 2140 | 1240 | 900 | 1190 | -750 | 1170 | 62 | 76 |
| 20 | 149 | 2950 | 2150 | 1200 | 950 | 386 | -800 | 378 | 60 | 80 |
| 21 | 152 | 3000 | 1000 | 0 | 1000 | 0 | -2000 !!!! | -1620 | 0 | 0 |

DISASTER !!!!
The human population has totally destroyed its resources of Bio-fuels.
Program Ends OK . . .

*Hidden advantages of the modelling process*

It would be a serious fallacy to assess the value of building a model purely in terms of the results (either tabulated or graphical) which it produces, as the exercise of building the model is often very beneficial in its own right. The model's concise and explicit implementation as a computer program produces an extremely rigorous and unambiguous statement of the assumptions, mathematical formulae and data which underlie it. The full extent of this specification can, of course, only be appreciated by someone who can understand the computer program but to those experts the degree of specification is immense and exact. Indeed, the act of modelling may help highlight inconsistencies in previous thinking: if the results from a model seem counter-intuitive it is easier to investigate the causes of this discrepancy (i.e. an unrealistic model or fallacious intuition!) by using or altering the model than to speculate without one. As an illustration of these hidden advantages I will outline the various stages of development of this example model.

The need for the model became apparent during discussions about a course of lectures 'Micro-computers in Environmental Research' to be given at Khartoum University and the concept and general form of the model (e.g. as summarised by Figure 1a) were agreed between the botanist and modeller over a cup of coffee. As a result of this the modeller spent an hour or so producing a flow diagram (e.g. as in Figure 4, then expanded to include some computer code) and thereby discovered some inconsistencies and illogicalities in the original 'word' model. It took three hours of detailed discussion with the botanist to decide on new relationships and assumptions which simply, but adequately, overcame those snags. Writing the computer program on the basis of this revised prescription took about two days, in the course of which it became apparent that one snag had not been properly overcome, and another one hour discussion was needed to resolve this. The model was then tested, a process which showed it was not easy to compile a fully comprehensive set of data all in the right units. Some minor revisions and improvement of the input and output of data and results took another day or so. This preliminary version of the model was then used as a teaching aid (Bacon, Callaghan and Lindley) with the following results:

i) the botanist who had helped devise it was somewhat surprised at the large effects caused by small changes to some parameters and, conversely by the small effects produced by large changes to others—expressing his own ideas using the precision of a computer model had been an instructive exercise;
ii) the use of a clear example of a computer model in biology was much more readily understood by the students than more theoretical presentations;
iii) the relevance of the model was easily demonstrated to the students by compiling a set of data appropriate to their country, the Sudan, and comparing its predictions with calculations made by their own forestry departments;
iv) in the course of (iii) it became apparent that a number of publications inadequately defined whether they were talking about 'energy needed by people' or 'energy used by people', which emphasised to the students the relationships between 'energy needs' and 'efficiencies of use' both in practice and as defined in the model;
v) some aspects of the original model were too complex, and a few others too simple and seriously unrealistic in a few (unusual) circumstances;
vi) the formats of the input and outputs needed modifying and improving to enable students and other non-specialists to more easily relate the management options to the values of the driving parameters and to present the outcome in a way which dynamically illustrated the process the computer code was modelling.

As a result of the above experience in using the model a series of modifications was agreed (involving a three hour discussion). Writing and testing the new program took about six man-days and required three discussions of an hour or so to clear up new snags which arose from the required modifications. The addition of the 'cartoon graphics' illustrated program took about three man-weeks.

*What use are models?*

The model here described was envisaged as a teaching aid and is still under development. However, as outlined above, it already provides a concise and explicit summary of concepts that are used in the field of 'Bio-energy' which can be used both as a focus of discussion with other experts, and to illustrate to bureaucrats and managers the likely effects of policies which they may not have thought through fully. Even in its present form the model offers considerable scope for experts to change its data and formulae and investigate the consequences of those alterations. To these extents it has already served a useful function. The model could be extended to provide a crude cost-benefit analysis of model strategies in economic terms, which might make it of greater use to managers.

Scientists who look at the model in any detail are bound to find a number of assumptions and approximations which we recognise are crude approximations to more detailed understanding. On this aspect I would like to make two points: First, while detailed understanding of some processes, such as primary production by photosynthesis, is available, the data to apply that detailed knowledge to, for example, a reed-swamp in tropical Africa, may often not exist. Second, incorporating such details as some scientists would like for particular instances would inevitably sacrifice the simplicity and generality which we explicitly built into the model for our pre-defined requirements. Anyone with different requirements needs a different model, but we hope ours might provide a useful starting point from which more specific (Precise and Realistic) models might be developed.

## ACKNOWLEDGEMENTS

I am grateful to Roy Harrison, a student from Lanchester polytechnic, who undertook the graphics programming as a self-teaching exercise.

## REFERENCES

Bacon, P.J.; Callaghan, T.V. and Lindley, D.K. (in press) An approach to Environmental Modelling with a case-study of Bio-Energy resources in the Sudan. Environmental Monographs Series, Institute of Environmental Studies, University of Khartoum, Sudan.

Holling, C.S. (Ed.) (1978) Adaptive Environmental Assessment and Management. Wiley, Chichester.

Jeffers, J.N.R. (1978) An introduction to Systems Analysis: with ecologial applications. Edward Arnold, London.

# NUTRIENT FLOW MODELLING

HUGH G. MILLER AND MICHAEL F. PROE

*Department of Forestry, University of Aberdeen,*
*and Macaulay Institute for Soil Research, Aberdeen*

SUMMARY

The importance of nutrient cycling was appreciated early in the study of forest nutrition, with perhaps the greatest stimulus being the fear that certain forest operations, including fire and timber harvesting, might lead to an irreparable loss of nutrients. Studies of full nutrient cycles had to await the advent of both automated chemical analysis techniques and computers to process the vast amount of data so obtained. By incorporating such data into models it has been possible to considerably extend our understanding of the nutritional requirements of a tree crop. Recently the relationships established by experiments have been incorporated into simulation models, which promise improved means of predicting the likely consequences of the different silvicultural options available to the manager.

INTRODUCTION

Man has probably had at least a rudimentary appreciation of nutrient cycling ever since he started deliberately to manure land. Real advance, however, had to await the development of modern concepts of plant nutrients. Concepts such as these develop slowly, but if a date has to be put to the start of our present understanding of plant nutrition it should probably by 1845 with the publication of the work of Liebig. He demonstrated that the soil was not the source of the carbon in plants but rather supplied salts and water. In 1885 in his 'Principles of Agricultural Chemistry' Liebig further pointed out that the action of what we now term nutrients was 'within certain limits directly proportional to the mass or quantity of these substances, and inversely proportional to the obstacles or the resistance which impede these actions'. This neatly encapsulates the twin concepts of amount and critical pathways or transformations that underly all nutrient cycling studies (Figure 1).

Forestry can claim a very early interest in nutrient cycling for in 1876 Ebermayer convincingly explained the poor health and declining growth of many Bavarian forests in terms of the export of nutrients in the leaf litter removed annually by local farmers. Following this, however, most work centred on agriculture. Interest in forest nutrition waned for it seemed observable that trees needed little nutritional assistance. Indeed, in his textbook of 1904 Schlich felt able to state that 'almost any soil can furnish a sufficient quantity of mineral substances for the production of a crop of trees . . .' He did, however, add the important caveat 'provided the leaf mould (humus) is not removed . . .'

Within a few decades of Schlich's pronouncement it was already clear that many of the soils on which new forests were being planted were unable to supply, unaided, sufficient amounts of all nutrients to ensure successful forest growth. Seventy-five

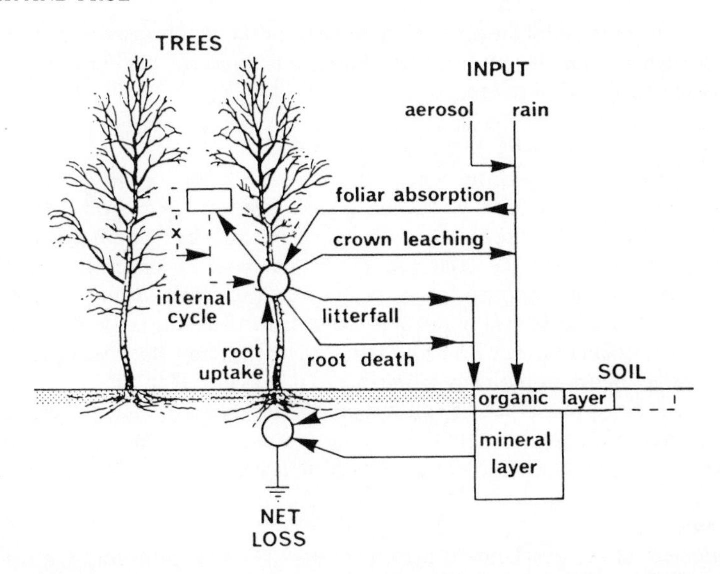


FIGURE 1. Illustration of the nutrient cycle in forests. Zones of accumulation are represented by rectangles and mobile pools by cirlces, loss through leaching to deeper soil horizons represented as loss to earth.

years after Schlich, Pritchett (1979) was to point out in his textbook that the 'correction of nutrient deficiences is common practice in pine forests of southern United States, in coniferous forests of the Pacific Northwest, central Europe and Scandinavia, and in exotic forests of Australia-New Zealand, South America and southern Africa'. This increasing need to resort to fertilizers at the establishment phase provoked worries about future requirements. Would there be a continuing need to fertilize or could the cycle of nutrients be progressively improved? Interest in nutrient cycling was further stimulated by realisation that increasing intensity of forestry may be accelerating the loss of nutrients.

Viro (1969) records that as early as 1930 the fear of nutrient loss was leading to a reduction in the use of fire in Finnish forests. Rennie (1955), on the basis of earlier literature values on the nutrient content of different tree components, came to the conclusion that in intensively managed forests on poor soils the rate of release of mineral nutrients may be inadequate to meet the rate of removal in harvested timber. Also about this time Ovington (eg 1962) started his important series of studies on forest nutrient cycling, initially stimulated by concern over the effects of fast growing tree species on soil development. Among other points he made was the fact that the nutrient input in rainwater may be sufficient to balance the drain in harvested timber that so concerned Rennie.

By the early 1960's it had become clear that it was dangerous to draw conclusions about the nutritional balance in forests, and the effect upon this of management operations, without a clear idea of the relative magnitude and interplay of the various pathways of nutrient flux. Studies of complete nutrient cycles, however, require the collection of large amounts of data (in excess of 10 000 analysis values in a recent study at the Macaulay Institute). Effective nutrient cycling studies, therefore, had to await the development of automated equipment for the rapid chemical analysis of a large number of samples and, no less important, powerful digital computers to process the results so obtained.

## PATTERNS AND PROCESSES IN NUTRIENT CYCLING

It has become convenient to represent nutrient cycling on three scales. The *geochemical* cycle introduces nutrients to and removes them from the forested ecosystem. This includes the input in rain (rain is far from being pure water) and the nutrients leached from the soil and hence removed in drainage water. The *biogeochemical* cycle comprises processes such as the uptake of nutrients by trees and their eventual release in litter fall or through crown leaching (Figure 1). The latter is the loss of nutrient elements from plant tissues into the rainwater passing over them, a particularly important pathway in the flux of potassium. Thus the biogeochemical cycle is the cycle through the tree-soil system. The *biochemical* cycle is then the cycle within the tree, including not only the movement of nutrients from roots to shoot but also the recovery and storage of nutrients from leaves as these die and are shed.

### *Geochemical cycles*

Measurement of the geochemical cycle presents particular problems. On the input side, rainwater is relatively easily collected, although there are some difficulties in analysing such a dilute solution. However, in addition to the input in rainwater, there is a further input as aerosols, dust and gases trapped from the atmosphere by impingement upon tree surfaces. This input has been termed the filter effect of trees or interception deposition. For sulphur and nitrogen some of this input is as gases, such as sulphur dioxide, oxides of nitrogen, ammonia and nitric acid gas, for which entrance through the stomata is probably a major part of the collecting process. Estimation of this input is a complex process involving detailed micrometeorological measurements and considerable computation (Bache, 1977; Fowler, 1980).

In principle it might seem easier to assess that portion of the interception deposition that arrives as aerosols, mostly in the form of mist and fine raindrops. Surrogate surfaces, such as netting, can be used to demonstrate the process. However, conversion of such results to amount collected on a hectare of trees is presenting a considerable problem. The difficulty lies essentially in developing means of partitioning the increased nutrient load in water collected beneath trees into that portion contributed by intercepted aerosols and that portion contributed by crown leaching. One approach has been to use the enrichment beneath leafless deciduous trees in winter as a measure of the aerosol input (Mayer and Ulrich, 1974; Miller, in press), but clearly this disregards the greater trapping efficiency of trees in full leaf. An alternative, initially proposed by Miller (Miller *et al.* 1976a; Lakhani and Miller, 1980), is to use the regression of the measured increase beneath trees on that beneath a surrogate surface (a fine mesh cylinder mounted above a rain gauge), and take as crown leaching the predicted increase beneath trees when input to the surrogate surface is zero. Again this is a process that requires the collection and computation of a large amount of data but, for sea-derived elements (particularly sea salt) at least, it can be quite effective. Recent developments, including further improvement by comparison of ionic ratios, have come from America (Lovett and Lindberg, in press) and New Zealand (Baker *et al.,* in press).

Loss from the ecosystem also presents problems of assessment because there is no

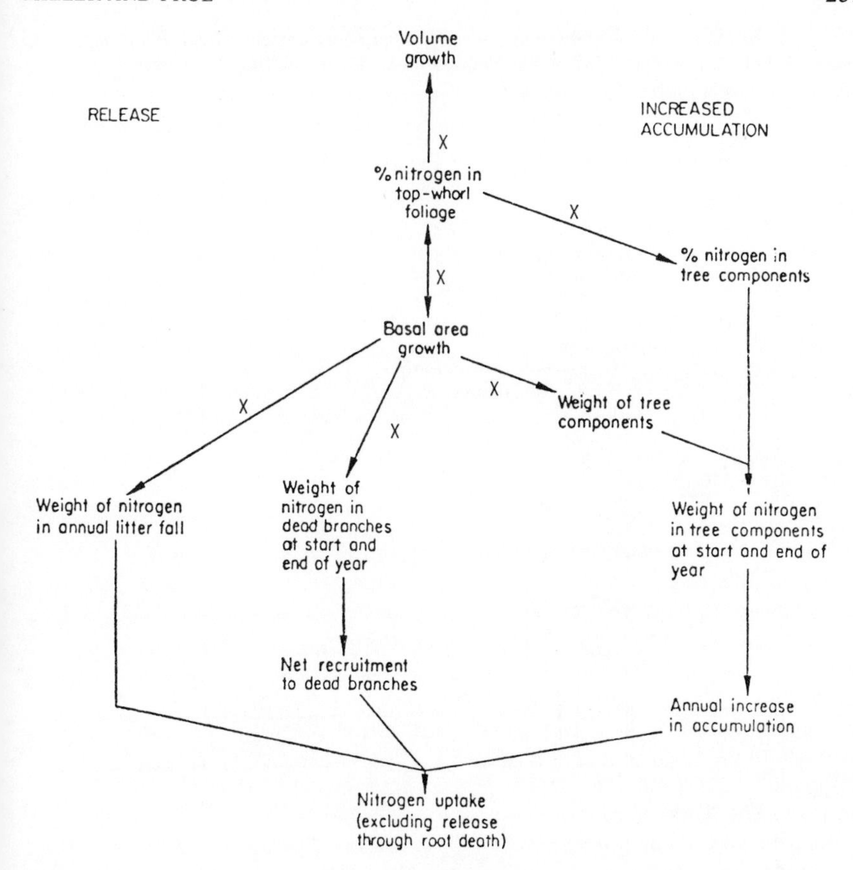


FIGURE 2. Flow chart used in calculating the nitrogen uptake required to maintain any volume growth in Corsican pine. Calculations based on empirically derived regression formulae marked x (Miller).

effective means of measuring the amounts of water draining through the soil, except on a catchment-wide basis. Most investigators, therefore, have had to resort to some form of hydrological model, of which the most sophisticated is that of Van der Ploeg *et al.* (1976).

*Biogeochemical cycles*

There are further measurement problems at the biogeochemical scale, the most obvious being that nutrient uptake by roots is not open to direct assessment. Estimation of this process, therefore, has to be obtained by summing measurements of the rate of nutrient accumulation by trees and the rate of release of nutrients from these trees (Figure 1). In order to relate uptake to different rates of tree growth Miller *et al.* (1976b) had to derive and then solve equations forming a complex network of relationships (Figure 2), a procedure that would be extremely laborious in the absence of computers. This technique, however, did produce answers to questions on the

TABLE 1. Comparisons of sinks and sources of nutrients calculated for *Pinus nigra* var. *maritima* of stand volume 230m$^3$ ha$^{-1}$ and volume increment 20m$^3$ ha$^{-1}$ year$^{-1}$.

| Flux | kg ha$^{-1}$ year$^{-1}$ | |
|---|---|---|
| | N | P |
| Sinks | | |
| To new needles | 92 | 9.4 |
| To structural tissue | 46 | 4.7 |
| Total requirements | 138 | 14.1 |
| Sources | | |
| Uptake from soil | 69 | 6.0 |
| Recovery from needle before death | 61 | 7.2 |
| Recovery from other tissues | 8 | 0.9 |
| Total of all sources | 138 | 14.1 |

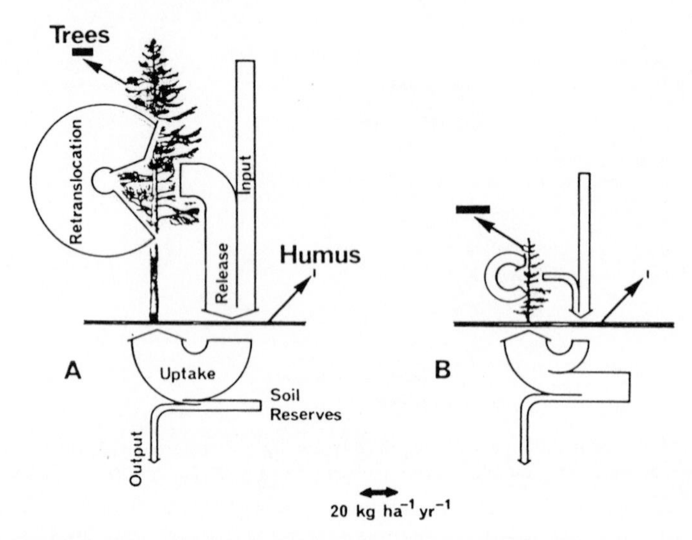


FIGURE 3. Patterns of cycling of potassium calculated (A) for 40-year-old and 11m tall pine and (B) for 10-year-old and *circa* 2m tall pine. Width of arrows are proportional to the fluxes they represent and the lengths of the solid bars are proportional to the rates of accumulation in trees and humus (Miller, in press, b).

interaction between the uptake, storage and subsequent re-use of fertilizer nitrogen. For example, it demonstarted that the longevity of fertilizer response is explainable in terms of nutrients stored within the tree tissue rather than in any acceleration in the biogeochemical cycle.

*Biochemical cycles*

The importance of the recovery, conservation and re-use of nutrients within the tree, i.e. the biochemical cycle, has been emphasised by further calculations (Table 1). These show that 44 percent or more of the nutrients required for new growth in closed-canopy forests is recovered from the previous year's dying needles. This immediately posed the question as to what would be the case in young trees which are

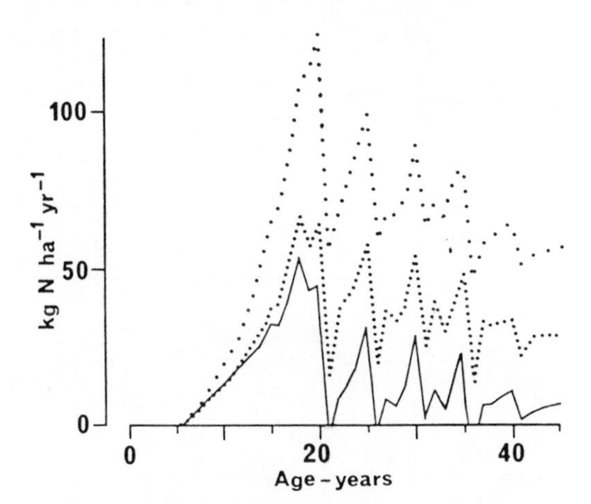


FIGURE 4. Modelled nitrogen requirements by age for Yield Class 6 birch (. . . . .) together with the remaining requirements after first allowing for recovery through retranslocation from dying foliage (. . . . . . . .) and secondly for the release on litter decomposition (________________). The remaining amount is the additional quantity of a nutrient that has to be incorporated into the cycle each year.

forming many more leaves that they are discarding because they are still expanding their crowns. Again it proved possible to seek an answer by constructing computer models based on the relationships already determined. By this means it was demonstrated (Figure 3) that the demand made on soil nutrient reserves by young pine is much greater than that by old pines. Thus the old trees represented as A (Figure 3) are taking from the soil less than 10 percent of the potassium required by the young trees represented as B. Given the understanding so acquired it then became possible to model the pattern of nutrient demands for other species, including alder (Miller, 1983) and birch (Miller, in press, a) based on data gleaned from the literature. For birch (Figure 4) the pattern of mineralisation from litter was included by calculating the release on decomposition for each year's litterfall, independently, by assuming an initial release rate in the first year of fall followed by a second rate to be applied to the amount remaining at the start of every subsequent year. The results so obtained for every cohort of litter were summed by years to give the total release rate. These simple though tedious calculations are ideally suited to computers.

## SIMULATION MODELLING

The models so far discussed were means of asking specific questions from given sets of data. Once enough is known about a system, however, it should be possible to construct a model that is sufficiently complete to be used to subsequently predict the answer to new questions as they arise. A genuinely predictive model is probably still a long way off, but nevertheless simulation modelling offers a means of encapsulating our present knowledge in a rigorous and amenable form. The predictions from such a model are only as good as the assumptions and empirical data that went into it, and this must never be forgotten. However, the same is true for all predictions, whether they come through a computer or not.

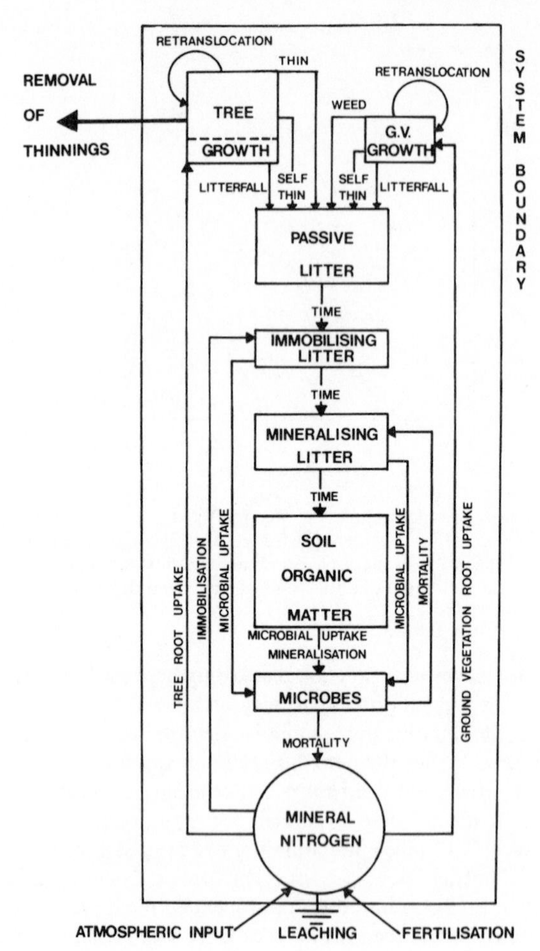


FIGURE 5. Elements of the nitrogen cycle simulation in FENDS.

It will be clear from the preceding discussion that nutrient modelling first requires a growth model as much of the cycle is bound up with the accumulation and transfer of biomass. Such a growth model can either be based on a knowledge of physiological processes or on empirical growth curves. Perhaps the best developed nutrient cycling model is FORCYTE, produced by Kimmins and Scouller (1983) in British Columbia. This they defined as 'an ecosystem-based forest management gaming model designed to examine the long-term consequences of intensive forest management on site nutrient capital, biomass production and the economic performance and energy efficiency of alternative management scenarios'. This model essentially uses yield curves to predict biomass accumulation, the pattern of which is then modified if the model finds that the nutrient supply is inadequate. As Kimmins and Scouller (1983) point out, they have aimed to produce a management tool whereas many models are designed to illustrate or explain aspects of fundamental scientific interest. The latter

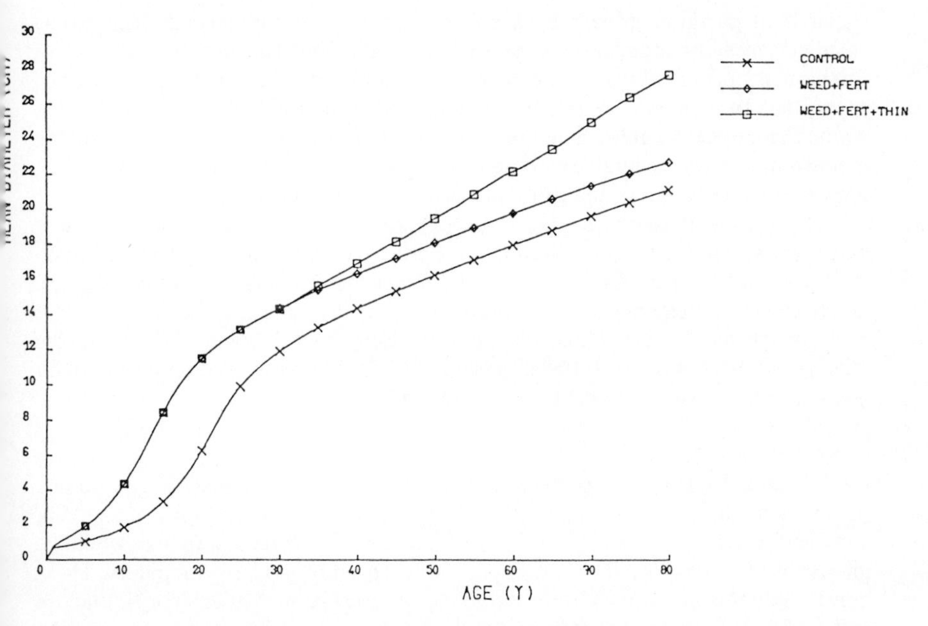


FIGURE 6. Response of tree mean diameter to various cultural operations as predicted for a particular site type by the simulation model FENDS.

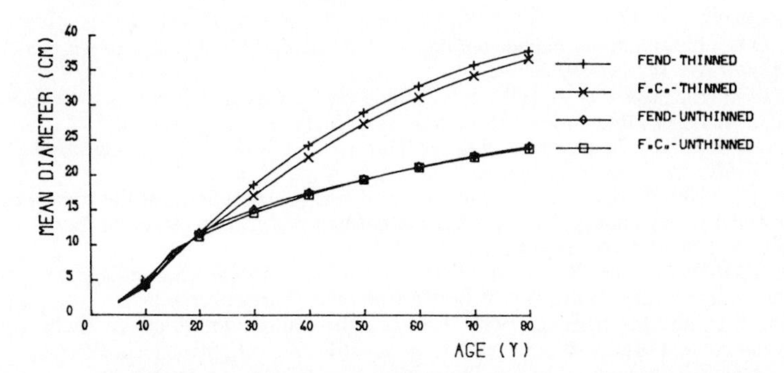


FIGURE 7. Mean diameters predicted by the simulation model FENDS in comparison to the equivalent values given in the appropriate Forest Commission Yield Tables.

require more detailed input than is ever likely to be available for routinely managed stands. However, it might be argued that basing a model on empirical yield curves reduces the ability to tailor the model to particular unique sets of factors applicable to each individual site as yield curves give generalised growth patterns.

Accordingly, in the FENDS simulation model being developed at Aberdeen (Proe, 1985), growth is based on the efficiency of foliage in converting light energy to biomass. This efficiency is assumed to vary with site factors, other than soil nutrient availability, while allowance has to be made for changes in light extinction between canopies of different densities. Elaborate equations, that vary with tree size, are then required to partition growth between the various tree components. The model calculates mensurational parameters, such as height and basal area, from the values for biomass accumulation. To achieve the calculated potential growth nitrogen supply has to be adequate, this being computed from a nitrogen cycle simulation within the model. An abbreviated representation of this is given in Figure 5. Such a model requires a considerable amount of computation but, as illustrated in Figure 6, does enable prediction of the consequences of various management options.

The precision of the model is illustrated in Figure 7 where predicted mean diameter values are compared to those given in the Forestry Commission Yield Tables. Such models are still in their infancy but with the increasing availability of computers and, in particular, as these become ever more 'user friendly', considerable advances are likely in the near future. Such advances are likely to be prompted more by the manager's wish to use the flexibility provided by ready access to computers than by any desire by research scientists to provide the models.

## CONCLUSION

It is now possible in many branches of science to collect exceptionally large amounts of data in a relatively short space of time. This presents the opportunity of studying complete systems such as nutrient cycles. Analyses of such data, however, would be impossible if it were not for the computational speed of modern computers. These now present the opportunity of reassembling cycles as mathematical models that are sufficiently realistic to enable the manager to assess the likely consequences of the different options open to him.

## REFERENCES

Bache, D.H. (1977) Sulphur dioxide uptake and the leaching of sulphates from a pine forest. *J. Appl. Ecol.* **14**, 881-895.

Baker, T.G.; Hodgkiss, P.D. and Oliver, G.R. (in press) Accession and cycling of elements in a coastal stand of *Pinus radiata* D. Don in New Zealand.

Ebermayer, E. (1876) *Die gesamte Lehre der Waldstreu mit Rüksicht auf die chemische Statik des Waldbaues.* J. Springer, Berlin.

Fowler, D. (1980) Removal of sulphur and nitrogen compounds from the atmosphere in rain and dry deposition, pp22-32, in *Ecological Impact of Acid Precipitation* (Eds D. Drabløs and A. Tollan), SNSF-project, Oslo-Ås.

Kimmins, J.P. and Scouller, K.A. (1983) *Forcyte 10, A User's Manual (Second Approximation).* Faculty of Forestry, University of British Columbia, Vancouver.

Lakhani, K.H. and Miller, H.G. (1980) Assessing the contribution of crown leaching to the element content of rainwater beneath trees, pp 161-172, in *Effect of Acid Precipitation on Terrestrial Ecosystems* (Eds T.C. Hutchinson and M. Havas). Plenum Pub. Corp., New York.

Liebig, J. (1885) *Principles of Agricultural Chemistry with Special Reference to the Late Researches made in England.*

Liebig, J. von (1845) Chemistry in its Applications to Agriculture and Physiology. Reprinted in Leith, H.F.H. (1978) *Patterns of Primary Production in the Biosphere,* Benchmark Papers in Ecology 8. Dowden, Hutchinson and Ross, Stroudsberg, Penn.

Lovatt, G.M. and Lindberg, S.E. (in press) Dry deposition and canopy exchange in a mixed oak forest as determined by analysis of throughfall.

Mayer, R. and Ulrich, G. (1974) Conclusions on the filtering actions of forests from ecosystems analysis. *Oecol. Plant.* **9**, 157-168.

Miller, H.G. (1983) Nutrient cycling in alder. *IEA Report NE 1983: 2.* National Swedish Board for Energy Source Development.

Miller, H.G. (in press, a) Nutrient cycles in birchwoods. To appear in Proc. Bot. Soc. Edinb., Birch Symposium.

Miller, H.G. (in press, b) Dynamics of nutrient cycling in plantation ecosystems. Chapter 2 in *Nutrition of Forest Trees in Plantations.* Academic Press, London.

Miller, H.G.; Cooper, J.M. and Miller, J.D. (1976a) Effect of nitrogen supply on nutrients in litter fall and crown leaching in a stand of Corsican pine. *J. Appl. Ecol.* **13,** 233-248.

Miller, H.G.; Miller J.D. and Pauline, O.J.L. (1976b) Effect of nitrogen supply on nutrient uptake in Corsican pine. *J. Appl. Ecol.* **13,** 955-966.

Ovington, J.D. (1962) Quantitative Ecology and the Woodland Ecosystem Concept. *Adv. in Ecol. Res.* **1,** 103-192.

Ploeg, R.R. van der; Beese, F. and Benecke, P. (1976) Simulationsmodelle von Wald-Ökosystemen: Wasser, pp 29-41 in *Verhandlungen der Gesellschaft für Ökologie,* Gottingen, 1976.

Pritchett, W.C. (1979) *Properties and Management of Forest Soils.* John Wiley and Sons, New York.

Proe, M.F. (1985) Unpub. PhD Thesis, University of Aberdeen.

Rennie, P.J. (1955) The uptake of nutrients by mature forest growth. *Pl. Soil* **7.** 49-95.

Schlich, W. (1904) *Schlich's Manual of Forestry, Vol. II, Silviculture.* Bradbury, Agnew and Co., London.

Viro, P.J. (1969) Prescribed burning in forestry. *Comm. Inst. For. Fenn.* **67**(7).

# EXPERT SYSTEMS: AN APPLICATION IN SOIL RESEARCH

P. INESON

*Institute of Terrestrial Ecology, Merlewood Research Station, Grange-over-Sands, Cumbria LA11 6JU*

SUMMARY

There are few computing areas which have aroused as much attention in recent years as expert systems. The adoption of knowledge-based systems as the basis for the 'fifth-generation' computer project in Japan is an indication of the growing importance of rule-based programming and, consequently, for expert system production. In the current paper an introduction to expert systems and their application is given. A worked example is given, showing how data analysis may be performed in order to produce rules with the generated rules then being applied in the construction of a simple expert system. The example chosen is taken from the field of soil science.

INTRODUCTION

'Expert systems' are becoming fashionable in computing circles, being heralded as the panacea for problems in medicine, molecular chemistry, ecology, business management, and many other fields of human expertise. This paper is intended to inform the forester about expert systems and how they work, and to give an indication of how they may assist him in his duties in the future. Different types of existing expert systems are presented, and a worked example given to explain how 'rule-based' programming offers an additional weapon in the computing arsenal.

An expert system can be loosely defined as a program that can replace a human expert in some field. More rigorously an expert system is defined as:

> the embodiment within a computer of a knowledge-based component from an expert skill in such a form that the system can offer intelligent advice or take an intelligent decision about a processing function. A desirable additional characteristic, which many would consider fundamental, is the capability of the system, on demand, to justify its own line of reasoning in a manner directly intelligible to the enquirer. The style adopted to attain these characteristics is 'rule-based programming' (Naylor, 1983).

Certain advantages of expert systems over human experts become immediately apparent, yet other less obvious advantages exist. Obviously, a computer expert can always be on call, is comparatively inexpensive to use, and can be copied as many times as it is needed. It is consistent and transparent, enabling the user to see why decisions have been reached and to be certain that the decision has not been affected by mood or temperament. Additionally, an expert system can acquire knowledge from several experts either within or between subject areas, surpassing the abilities of the single individual. Unlike a human expert, an expert system will search through possibilities often too numerous or too counter-intuitive for the human expert to deal with. Perhaps the most important of all is the need for the expert to formalise his knowledge into definite rules, thus making information available which has never previously been explicit.

*Examples of expert systems*

One of the first expert systems was created at Stanford University in 1965. This was developed by computer scientists and chemists who sought to automate the process of deducing correct chemical structures from basic analytical and spectrometric data. It was argued at Stanford that if the lines of reasoning, or the rules, used by the expert in deducing the correct chemical structure of a molecule could be formalised, then it would be possible to programme a computer to carry out the same process. The system resulting from that research is called DENDRAL and this has been used on hundreds of occasions to successfully predict molecular structures (Lindsay *et al.*, 1980).

Other successful examples of expert systems include PROSPECTOR, which is a system which interprets soil and geological data in a probabilistic fashion in order to locate mineral deposits. It has proved extremely useful and has predicted the location of a molybdenum deposit estimated to be worth $100 million. This has, not surprisingly, spurred the development of many industrially-oriented expert systems prospectors (Hayes-Roth *et al.,* 1983).

In ecology and forestry there are, as yet, few examples to cite, and some of these stretch the limits of an expert system. However, two examples help to show the potential:

> The problems of intoxication due to people experimenting with eating wild mushrooms and toadstools has led to a recent increase in the number of cases brought to the attention of the medical profession. The medical treatments differ between different ingested basidiocarps, yet it takes an expert to identify fungi from fragments and circumstantial evidence. Margot *et al.* (1984) have produced an on-line system for the identification of 224 poisonous mushrooms, calling on the minimum of specialist knowledge. The output lists the names of likely species and also appropriate medical treatment. The system can be run on a micro-computer, and has effectively distilled the knowledge of experts into a form which can be consulted by an educated layman.
>
> Starfield and Bleloch (1983) presented a simple expert system to advise on the management of areas of the Kruger National Park. The decision to burn blocks of land on the basaltic region of the Kruger National Park is dependent on several key points: history of burning; density of certain floral species; types of grazers; etc. These decisions are traditionally made on the basis of qualitative data and an expert's accumulated experience, yet Starfield and Bleloch have transferred this knowledge into a expert system. The system is transparent and makes the experts' knowledge explicit.

*Construction of an expert system*

The construction of an expert system requires certain common steps which can be achieved in a variety of ways. The steps are:

a) preparation of a knowledge base which contains the rules (the rules being provided directly by experts or by induction from a set of data) and
b) embodying the rules into a computer program.

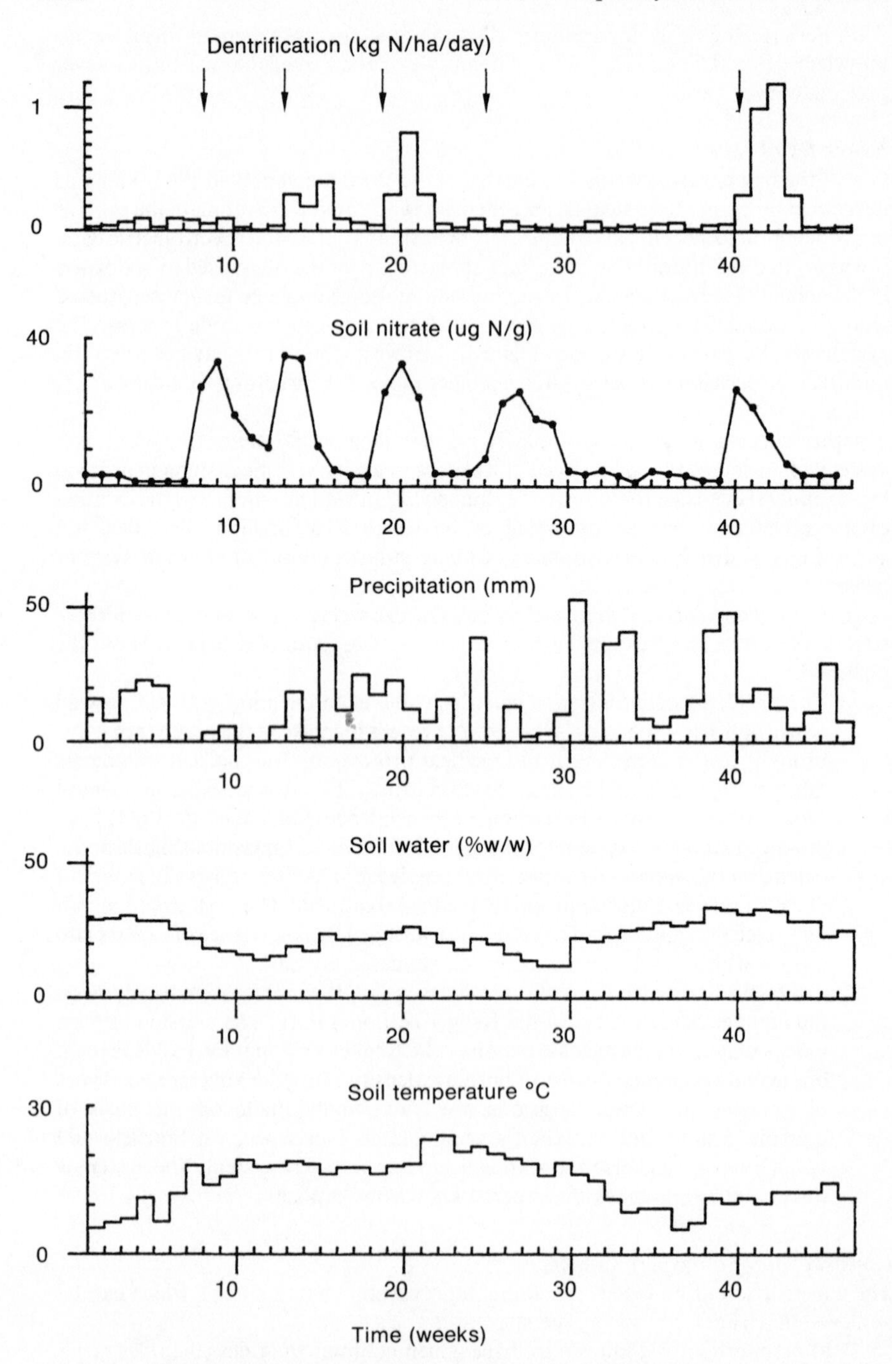


FIGURE 1. Diagram showing the data from Ryden (1983) analused using the HULK expert system. The data are from an experimental plot in Berkshire and show the relationship between soil dentrification rate and environmental variables.

FIGURE 2. Data from Figure 1 prepared in a format suitable for inputting into HULK.

```
 10 DENITR,    46,  5,
 20 DENIT
 30 NITRATE
 40 RAINFALL
 50 MOISTURE
 60 TEMP
 70 WEEK 1,    6, 30, 159, 289,  53
 80 WEEK 2,    4, 30,  77, 293,  64
 90 WEEK 3,    7, 30, 189, 304,  72
100 WEEK 4,   10, 12, 228, 285, 114
110 WEEK 5,    3, 12, 208, 285,  66
120 WEEK 6,   10, 12,   0, 254, 123
130 .
140 .
150 .
520 WEEK 46,   4, 32,  79, 251, 114
```

A simple worked example from the author's area of research is used here, with the objective of producing an expert system which can predict when nitrogen will be lost as gas (denitrification) from a specific soil in Berkshire. The data are taken from Ryden (1983), and are shown graphically in Figure 1. The aim of the exercise is to produce an expert system which can predict under what environmental conditions (levels of soil moisture, soil nitrate, temperature, rainfall rates) the process of denitrification will occur to any significant extent at this specific site.

A more conventional approach to analysing such a data set is to use multiple regression, with denitrification as the dependent variable, and soil moisture, nitrate content, temperature and rainfall as the independent variables. Such an approach has been used with the current data set and does little to explain the rates of denitrification, with only 50.1 percent of the variance being accounted for.

The first stage in the rule-based approach adopted for the current study was to translate the data into machine-readable form from which rules could then be deduced. The process of rule inference can be carried out by several commercially available packages or by user-written software (see Naylor (1983) for example). I have used one of the less expensive and more rapidly available systems called HULK (Helps Uncover Latent Knowledge) (Warm Boot Ltd., 40 Bowling Green Lane, London EC1R0NE) which runs on a BBC Model B micro-computer, and compares favourably with systems many times more expensive (Naylor, 1984). HULK does not discover rules *per se* but makes it very easy to test out rules which the user considers to be important, and provides a suitable vehicle for demonstrating how rules may be derived. HULK requires the data to be in the form of a line-numbered BASIC program. Figure 2 shows the data from Figure 1 prepared in a format suitable for input into HULK. The first line identifies the name to be given to the data set, the number of cases and variables. This is followed by a list of variable names and the actual data array. The data have been transformed to integer values prior to calculation (thus denitrification rates are multiplied by a factor of 10).

The initial phase in deriving rules from the data is to screen coarsely the variables to identify those which appear linked to the dependent variable under investigation. Visual inspection of the data suggests that significant denitrification occurs when losses are in excess of 2 kg N $ha^{-1}$ $d^{-1}$ (or 20 after integer scaling). We therefore test to see which variables are associated with the conditions (termed 'hypothesis' by HULK) of daily denitrification rates (DENIT)>20. Figure 3 shows the output from a run of Prescan, the component of HULK designed to perform this coarse screen. Prescan has effectively worked through each variable in turn, determining which, if any, are of use in explaining the 'hypothesis' DENIT>20. This is achieved by performing a simple approximate t-test on each variable in turn, with the values for the variable grouped into two classes, depending on whether the corresponding value for DENIT is less than or greater than 20. The results suggest that the variables nitrate and soil moisture would be worth pursuing.

The next program, called Look, asks for rules to test against the 'hypothesis' and then advise whether the rule should be kept, or abandoned. Rules can be added to form quite complicated Boolean expressions, and in the current example the following rule was tested:

DENIT>20WHEN NITRATE>60AND MOISTURE>190AND RAINFALL>0

The effectiveness of the rule can be assessed using the final program, called Leap, which tests the rule against each case in the data set and measures the success of the rule in satisfying the 'hypothesis.' In the current example, 100 percent success was achieved, and so the rule outlined above explained the occurrence of denitrification at the site under study (Figure 4).

The final stage in the production of an expert system is to convert the rule(s) into a program which can successfully interact with non-experts. In the current example the rule which has resulted from the analysis by HULK has been converted into a simple BASIC program which runs on a BBC microcomputer. In response to an input of soil nitrate content, soil percentage moisture and rainfall data the program simply uses the rules presented above to decide whether denitrification will exceed 2.0 kg N/ha/day. When asked 'why' the program responds with the reason why denitrification will or will not occur. The information present in Ryden (1983) has thus been used to produce a simple expert system capable of making specific predictions about the occurrence of denitrification at the site under study. This system, on its own, is of little practical importance, yet when coupled with similar 'experts' on nitrogen leaching and ammonium volatilisation could provide a means of advising the farmer when to apply nitrogen fertilisers to minimise losses.

In this example a particular data set has been used to derive general rules which after testing can be incorporated into a decision-making program. In the construction of expert systems the rules can be obtained in several ways, either from raw data as shown here, from existing human experts, or from the literature. The process of attempting to derive rules has resulted, in the current example, in a greater understanding of the data than was achieved through more conventional statistical techniques. Also, this approach has the advantage that the results of the analysis are in a form which is more easily explicable to the non-scientist.

The power of the modern computer, coupled with the sophistication of the new generation of rule-based programming languages, enables the collection and interaction of rules to build systems of enormous complexity. In the field of ecology, where we are attempting to understand complex systems, the construction of such models will greatly facilitate our understanding of how natural systems function. However, it will become increasingly important for environmental scientists to present the results of their research in a manner which permits incorporation into systems of this kind.

FIGURE 3. Output from Prescan for the data input in the form shown in Figure 1.

```
HYPOTHESIS: denit>20
        Variable 2              nitrate
                          SUCCESS 9           FAILURE 37
            average:  217.89                  81.59
            std. dev.:  98.05                 95.23
            difference-score=3.76**

        Variable 3              rainfall
                          SUCCESS 9           FAILURE 37
            average:  167.11                  148.05
            std. dev.:  91.74                 149.27
            difference-score=0.49

        Variable 4              moisture
                          SUCCESS 9           FAILURE 37
            average:  270.11                  225.95
            std. dev.:  46.36                 57.91
            difference-score=2.43*

        Variable 5              temp
                          SUCCESS 9           FAILURE 37
            average:  157.44                  141.11
            std. dev.:  42.39                 51.15
            difference-score=0.99

        Date from DENDATA
        NB High difference-score indicates a useful variable
```

FIGURE 4. Output from the Leap sysyem of HULK showing the success of the generated rule in satisfying the hypothesis.

```
HYPOTHESIS IS
    denit>20

Rule 1: nitrate>60 AND moisture>190 AND rainfall>0

SAMPLE              SUCCESS
 1 *WEEK 41         1
 2 *WEEK 40         1
 3 *WEEK 21         1
 4 *WEEK 43         1
 5 *WEEK 42         1
 6 *WEEK 13         1
 7 *WEEK 19         1
.
.
.
46 *WEEK 1          1

Date from DENIT
Rules used: FINRUL

Success rate=100%
```

REFERENCE

Hayes-Roth, F.; Waterman, D.A. and Lenat, D.B. (1983) *Building expert systems.* Addison-Wesley, Massachusetts.

Lindsay, R.K.; Buchanan,B.G.; Feigenbaum, E.A. and Lederberg, J. (1980) *Applications of artificial intelligence for organic chemistry: the* DENDRAL *project.* McGraw-Hill, New York.

Margot, P.; Farquhar, G. and Watling, R. (1984) Identification of toxic mushrooms and toadstools (Agarics)—A on-line identification program. In: *Databases in Systematics.* pp 249-261. Eds. R. Allkin and F.A. Bisby. Academic Press, London.

Naylor, C.M. (1983) *Build your own expert system.* Sigma Technical Press, Wilmslow.

Naylor, C.M. (1984) Discriminating experts. *Practical Computing* 7(3), 108-112.

Ryden, J.C. (1983) Denitrification loss from a grassland soil in the field receiving different rates of nitrogen as ammonium nitrate. *Journal of Soil Science* **34**, 355-365.

Starfield, A.M. and Bleloch, A.L. (1983) Expert systems: an approach to problems in ecological management that are difficult to quantify. *Journal of Environmental Management* **16**, 261-268.

# INTELLIGENT KNOWLEDGE-BASED SYSTEMS: POTENTIAL FOR FORESTRY

ROBERT MUETZELFELDT

*Department of Forestry and Natural Resources, University of Edinburgh, Kings Building, Mayfield Road, Edinburgh EH9 3JU*

## SUMMARY

Recent developments in artificial intelligence have opened up the prospect of computer systems which can mimic the abilities of human experts. Such systems will store knowledge about the real world as facts and rules, and will be able to reason using this knowledge. In forestry, these systems will give managers access to the accumulated wisdom of forestry expertise. They will behave as 'intelligent assistants', informing, advising and explaining, in all aspects of the complex task of managing a forest.

## INTRODUCTION

The Japanese Fifth Generation Computer Project aims to produce within ten years computers which will contain knowledge about the real world, and to surpass the intellectual abilities of man in using this knowledge (Feigenbaum and McCorduck, 1984), Other countries are responding to this initiative, realising that to lose out in this race could mean relegation to the second division in the world economic league. In Britain, for example, the Alvey Programme is putting $200 million into advanced information technology: one of its four areas is the development of intelligent knowledge-based systems (IKBS).

As with many other disciplines, forestry involves the processing of a great deal of information. This is not just a data-processing activity: any forest manager will bring in a good deal of implicit knowledge about the real world, awareness of local conditions, and balancing of issues, to arrive at his decision: in other words, he will use his knowledge in a 'intelligent' manner. It is therefore relevant to contemplate the development of computer systems that can capture knowledge in the field of forestry, and which can use the knowledge in ways that we would call 'intelligent'.

In addition, many of the sources of knowledge that a forester currently needs to consult are widely dispersed, amongst different experts or in various publications. An intelligent assistant would provide a single, integrating route into this diversity of knowledge.

In this paper I will discuss the nature and potential application of IKBS in forestry. We will consider an IKBS which takes the form of an intelligent (or rather, super-intelligent) assistant, sitting on your desk, and knowing 'everything there is to know' about forestry. How will knowledge be stored in such an assistant? What basic operations will it need to be able to perform on this knowledge? And what uses will we make of such an assistant?

REPRESENTING KNOWLEDGE ABOUT FORESTRY

One method for representing knowledge is based on the assumption that most knowledge can be expressed in terms of FACTS and RULES. The conventions used here for illustrating this approach are loosely based on the Prolog programming language, which in turn is derived from a form of logic called predicate logic (Clocksin and Mellish, 1981). The particular examples of facts and rules given below, in the context of insect damage, should be viewed as a tiny part of the knowledge held by the intelligent assistant, which would cover many other facets of forestry as well.

*Facts*

Facts specify the relationship between particular objects. For example, the following fact

*pine beauty can attack lodgepole pine* (Fact 1)

expresses the 'can attack' relationship between a particular insect species and a particular tree species, while

*fenitrothion can kill pine beauty* (Fact 2)

represents the 'can kill' relationship between an insecticide and an insect species. These facts are generally true across the whole field of forestry. In contrast,

*compartment 125 is next to compartment 127* (Fact 3)

is only true for a particular estate, while

*compartment 125 has species lodgepole pine* (Fact 4)

is only true for the present rotation: clear-felling and re-planting would requires this fact to be replaced by a new one specifying the species planted.

*compartment 127 has outbreak of pine beauty* (Fact 5)

is an example of a fact representing very changeable knowledge about the real world. Thus, we see that facts can express knowledge ranging from the generally-true to that which applies only at a particular point in time or space.

*Rules*

A rule is a general statement about the relationship between objects, rather than a specific statement given by a fact. A rule normally saves us having to state many facts.

For example, we can define the conditions under which a compartment is at risk from insect attack by the following rule:

*compartment C is at risk from insect I* IF (Rule 1)
*compartment C has species S* AND
*insect I attacks species S* AND
*compartment C is next to compartment C1* AND
*compartment C1 has outbreak of insect I.*

This says that a particular compartment is at risk from some insect if the compartment contains a species which the insect attacks, and if there is an outbreak of the insect in a neighbouring compartment.

Note that this rule saves us having to provide a long list of compartments at risk, but, more importantly, provides a basis for deducing, from other information, whether a compartment is at risk.

Rules can also be used to represent the conditions controlling management decisions, as in:

*spray compartment C with pesticide P* IF (Rule 2)
*compartment C is at risk from insect I* AND
*pesticide P kills insect I.*

which states that we should spray a compartment with some pesticide if the compartment is at risk from insect attack, and the pesticide kills the threatening insect.

## USING THE KNOWLEDGE

### *Asking questions: goal-directed inference*

Suppose we wanted to know what species attack lodgepole pine. We can view the request as partially filling in a template (a 'goal') which the computer could then apply to each fact or rule in the knowledge base, looking for a match. In this case, the template would look like:

*attacks lodgepole pine?*

and we would obtain a match with the first fact, with the blanks being replaced by pine beauty. So, 'pine beauty' would be the answer to our question. Note that, in answering this question, the computer does not need to understand English, but is simply involved in a pattern-matching operation.

Similarly, we could ask what pesticide we should spray compartment 125 with:

*spray compartment 125 with pesticide ?*

This time, the template would match rule 2, with C being set to 125, but with P not yet determined. Rule 2 would then itself generate two additional requests. The first would match rule 1, which would itself repeat the process; while the second would (in this particular case) match fact 2. The result of this process would be to replace the blanks by 'fenitrothion', this being the pesticide that kills the insect threatening compartment 125.

If the intelligent assistant is unable to answer some question because it lacks some information, then it should have the ability to come back to the user and ask for the relevant information. For example, if it does not know about the proximity of two compartments, then the user could be asked to provide this item. Thus, it is not necessary to ensure that everything needed to solve some problem is known in advance.

### *Volunteering information: data-driven inference*

The set of facts and rules can just as well be used in the reverse direction, with some conclusion being generated as soon as it can be deduced from a new bit of information. For example, on being told fact 5 (that there is an outbreak of pine beauty in compartment 127), the intelligent assistant could automtically deduce that compartment 125 is at risk from pine beauty, using rule 1, and then, using rule 2, that this compartment should be sprayed with fenitrothion. Thus, the intelligent assistant could continuously monitor events, and issue the appropriate warnings when something untoward happens.

### *Providing explanations*

We need to be able to question the basis for the actions taken by the intelligent assistant. There are two main types of explanation that we could seek:

HOW? (meaning: how did you arrive at this conclusion?) . . .

The intelligent assistant could print out the chain of rules it used to arrive at some conclusion.

WHY? (meaning: why do you want to know this item of information?) . . . The user would then be told how the item of information that he is being asked to provide could help the intelligent assistant in its reasoning process.

*Dealing with Uncertainty*

Often our 'facts' and 'rules' cannot be expressed with certainty. For example, we might wish to say that pine beauty IS QUITE LIKELY TO attack lodgepole pine; or that, if the conditions for Rule 1 are found to be true, that there is STRONG POSSIBILITY (but not a certainty) that the compartment is at risk. The phrases 'quite likely' and 'strong possibility' are expressed as probability levels (though in practice it is of dubious validity to attempt to quantify such objective notions of uncertainty). Our intelligent assistant must be able to reason with this uncertain knowledge to produce conclusions which can be stated with more or less certainty. This reasoning process might make use of established mathematical techniques for the analysis of risk.

*Incorporating mathematical models into the knowledge base*

If we wished our assistant to deduce the yield we could expect to get from our forest compartments, then we could provide it with a large number of rules which capture the information contained in, for example, the Forestry Commission Yield Tables.

However, there are now simulation models of forest growth which aim to provide greater accuracy under a wider range of conditions than that of the empirical tables (see the paper by Rennolls and Rollinson in these Proceedings). There is no problem here: we simply dress up the simulation model so that it appears as just another rule to the intelligent assistant, but actually initiates the running of a simulation model when it is invoked. Thus, our knowledge can be captured both as simple rules, and as a model representing stand dynamics: the intelligent assistant uses either as appropriate.

*Acquiring knowledge*

How does the intelligent assistant build up its store of knowledge?

One method is for it simply to be told facts and rules, just as the human expert can be taught during his training, or through experience. A problem here is to ensure internal consistency: different experts, adding knowledge to the system on different occassions, might make contradictory statements. Thus, there is a need to vet new information, and to check for clashes when reasoning over the knowledge base. There are also problems involved in deciding what level of detail to use for stored knowledge.

Another method (described by Ineson in these Proceedings) is that of induction: deriving rules from a collection of observations. For example, if a certain insect species was recorded as attacking a number of tree species, and if it were known that these species were all conifers, then the general rule could be induced that the insect will attack any species that is a conifer. This 'rule' of course must be subject to revision if contradictory information should later appear.

## USES FOR AN INTELLIGENT ASSISTANT

### *Extracting information from very large data bases*

The information held by the intelligent assistant will be of value in its own right. We can think, for example, of information on the spatial distribution of soil types and climatic variables; the characteristics of tree species and provenances, and all plants and animals that interact with forest tree species; yield tables; and details of weed control; and still have hardly begun. There will be many links between these various types of information, so it should be available all the time. For example, a forester interested in Norway spruce could want information on such diverse aspects as its susceptibility to wind-throw, its preferred silviculture conditions, economic aspects, and projected demand for Christmas trees.

A data-base system like this will inevitably need to be able to reason in satisfying the forester's requests. For example, it will need the ability of reasoning over taxonomic classification of plants and animals, to enable it to realise that information on voles is relevant to a request on rodent damage to young trees. It will also need to have abstract spatial knowledge, permitting it to work out the distance between a particular compartment and a saw-mill, knowing the distance of the road segments on the shortest route; and allowing it to answer questions about 'the compartments which are next to the compartments that border the river'.

Such systems will also need to guide the forester through the mass of information available, rather than passively waiting for the forester to enter a well-formulated request.

### *Expert systems*

Currently, the term 'expert system' is used to refer to computer systems that use a process of reasoning over a set of rules to solve a specific problem. For example, there are expert systems to identify infectious diseases; identify molecular structures; and evaluate sites for potential mineral deposits. Our intelligent assistant would in effect act as many expert systems rolled into one. The example facts and rules given in this paper could be seen as a part of an expert system advising on the use of insecticides to control forest insect pests. In a similar manner, we could extend the set of facts and rules to deal with the identification of a pest from symptoms of damage; guidance on an appropriate silvicultural regime according to local site conditions; and advice on the maintenance of diversity to aid in the conservation of wildlife species.

### *Giving access to powerful analytical facilities*

There is a very wide range of techniques in statistics, finance, operations research and modelling which a forester could profitably use, if only he had an easy way of accessing them. An intelligent assistant could act as go-between, asking, in a way the forester can understand, for the information needed to set up the analysis, re-arranging this in a form suitable for the computer program, then interpreting the results in the forester's language. As an example, a forester might wish to use a linear programming program to assist in the allocation of manpower and machinery, but not have sufficient knowledge of the technique, awareness of the constraints on its use, or familiarity with the details of the particular package to use it without assistance. An intelligent assistant could remove these barriers.

## CONCLUSIONS

This paper has, deliberately, given a very personal view of the potential for IKBS in forestry. On the one hand, I have simplified many of the issues surrounding the way that knowledge can be represented and reasoned with. Current research in Artificial Intelligence is exploring many alternative approaches to the one presented here. For many particular problems, these other methods might be more expressive or more efficient.

On the other hand, the picture I have painted of an intelligent assistant which has wide-ranging knowledge about forestry, and which can turn its hand to solving many different problems, goes considerably further than the IKBS developed to date. Most of these are designed for solving a very particular problem: an expert system for identifying a particular class of disease; an intelligent interface to a particular computer package. There is little doubt that the success of many of these systems is that they do deal with a restricted class of problem, and therefore can contain features specific to the class of problem which allow them to operate efficiently. There are certain to be many difficult issues to resolve in advancing from these specific systems to the general-purpose intelligent assistant postulated here; but there is little doubt that they will come, and will be of great benefit when they do.

## ACKNOWLEDGEMENTS

I would like to thank Dave Robertson and Dave Plummer for their constructive comments.

## REFERENCES

Clocksin, W.F. and Mellish, C.S. (1981) *Programming in Prolog*. Springer-Verlag.
Feigenbaum, E.A. and McCorduck, P. (1984) *The Fifth Generation*. Pan Books.

# PART VII

# SUMMING-UP

# SUMMING-UP—LESSONS LEARNED

MARK R. LEMBERSKY

*Group Information Systems and Finance Directory, (formerly Forestry and Timber Products R&D Director), Weyerhaeuser Company, USA*

Thank you, Mr President. It is kind of you to ask me to give some summary comments on the Conference.

Since we began on Tuesday and continuing through this morning, I have listened to the formal presentations, the questions following the presentation, and the informal discussions both during coffee breaks and in the evening. I have attempted to extract the 'lessons learned' that you and I might carry away from this Conference. The lessons I offer hopefully complement and supplement those conclusions each of you have already noted. Nearly all of the points that I will make either have been made explicitly somewhere in this Conference or, I believe, can be justifiably inferred from the formal and informal discussions that have taken place here. Of course, I bear responsibility for the interpretation that I will give you—and I have to admit that my own experience and biases will have some impact on my comments.

Let me begin with some broad generalisations:

1. The first is that we can conclude from this Conference that computers *can* be applied today in just about every aspect of forestry. This is obvious from the formal talks, from the demonstrations and from the trade exhibits. I am sure it is consistent with your informal discussions with other participants about activities in their organisations.

2. More important than the lesson that computers *can* be applied today is that this Conference has demonstrated very forcefully that computers *are* being exploited today in most areas of forestry. Computing in forestry is not pie in the sky. It is not something yet to come. We have seen at this Conference that *today* persons and organisations are gaining advantages by utilising computers in forestry. They are gaining an advantage for their organisation, gaining an advantage relative to their competition, and gaining an advantage for themselves professionally.

3. Further, again more important than the facts that computers *can* be applied today and that they *are* being exploited today is that *there is not any choice but* to utilise computers in forestry. Whether you and I like it or not, whether you and I think it should happen or not, western society is entering an Information Age that is the successor to the Industrial Age. Computers are not 'icing on the cake' of the industrial age, but are a harbinger of the new information age just as certainly as the cotton gin was not just a way to do agriculture better, but signalled the beginning of the industrial age.

This movement into a new information age is evident by the existence in the United Kingdom of a top-level ministry responsible for information technology, and reinforced by the presence of Minister Pattie at this Conference. Both his formal remarks and informal conversation at the dinner table make it clear that he views a significant part of his ministry's mission to be aiding the United Kingdom's transition from the industrial age into the information age.

In this dawning information age, dealing with computing is a necessity. It is a necessity for any organisation that hopes to survive economically and politically, and it is a necessity for any individual who hopes to prosper professionally. Now, this does not imply that all of us must be computer experts with considerable knowledge of bits and bytes and of hardware and software. But it does mean that we need to be ready to aggressively incorporate computing into our professional forestry lives. Indeed, as was said earlier in this Conference, a few years from today we will look back and wonder how the use or non-use of computers in forestry even could have been an issue.

So, whether one views computers in forestry as an offensive tool—that is, as a means to move ahead—or in a defensive manner—that is, as a necessity to avoid falling behind—certainly one lesson that we take away from this Conference is that computers in forestry are here to stay.

4. At the same time, there is no need for any individual or organisation to feel left behind in the utilisation of the computers in forestry. This is a result of the unique nature of computing technology. It is not necessary to retrace the paths of others who have already made headway in utilising computers. Because of the rapidly escalating power of hardware and software and the accompanying significant cost reductions, it is not necessary to make a large financial expenditure to move to the state-of-the-art. Also, there is no need to slowly move up the learning curve—rather, it is possible to leap-frog to near the top. One can buy a personal computer plus one or two good software packages and you can be up-to-date. There is no need for big expenditures, no need for long start-up times, no need to build up an experience base of years. In fact, a history of spending a lot of money, going through a lot of start up, and acquiring years of experience does not really give you that much of an advantage. Let me give you a personal illustration. In 'the old days' when I was doing 'real work' versus the managing I do now, I was relatively proficient in several computer languages. Today, I have a personal computer on my desk that is in constant use; however, I have yet to use any of those computer languages on my personal computer. I buy packaged software and have yet to use any of the old programs I wrote years ago. It simply is not necessary to have a prior experience in order to be a user at the forefront today.

Let us turn to what this Conference has had to say about hardware and software.

5. Let us first look at the lessons on hardware. I believe the fundamental lesson is that cost is rapidly coming down at the same time that hardware capability is equally rapidly going up. We, as users of this technology, are witnessing something that is as close to the 'free lunch' as we are likely to see. I hasten to add that while this is true for users of the technology, it is certainly not the case for those in the hardware business. The extraordinary competitive computer hardware environment is anything but a free lunch for those producing hardware.

During this Conference we have seen any number of concrete examples of software that is running on hardware costing today no more that £500 to £3 000 and that would have required hardware costing 10 or 100 times as much just a few years ago. Just to cite a few: Sven Westerling showed up his planning system running on a personal computer. The Forestry Commission told us about several data capture applications on hand-held computers. Last night Alan Moss demonstrated his forest management software that runs on a portable computer. By going through the Conference program we can pick out many more.

We heard some projections this morning of where hardware is heading in the future. I think it is fair to conclude that if for a particular application you cannot find hardware today that will do the job economically, just wait until 'tomorrow'. That is to say, do not put off getting ready for the application. Do not put off planning the application or gathering the data. It is very likely that by the time you carry out the planning and gather the data, the appropriate piece of hardware at a more economical price will be available.

7. What about which hardware to buy? Actually, there has not been a lot of discussion at this Conference related to brand name selection and I do not plan to endorse any particular brand. However, I will share with you my view on what to do when faced with a purchase choice between i) a standard widely used piece of hardware, and ii) the 'best hardware', meaning best for the particular application at hand because of some advantages in cost, performance, or special features. My recommendation is that you choose the *standard* hardware. It will have more software available for it. It is more likely to enjoy long-term suppport. Persons in your organisation will be more familiar with that hardware and the software that runs on it. The standard hardware is more likely to be acceptable to the persons in your organisation that ultimately authorise the purchase of the machine; a large corporation with a well known name is simply more comfortable for management than a less well known, less well established company. Further, there is a much lower risk that the firm producing the standard hardware will disappear. There have been many examples over the last few months of hardware firms failing or otherwise leaving the business and leaving owners of their products unsupported. Finally, the advantage that the best hardware has is likely to be only temporary. In another six months or so some other piece of hardware will come out that is superior. Then the currently best hardware will have lost its advantage and it will still have its advantages relative to the standard hardware.

Let us move to lessons we can draw regarding software.

8. When talking about software, it is important to differentiate between custom developed software—that is, software we write ourselves in-house—and packaged software that we buy. One of the lessons we learned at this Conference is that software we write ourselves in very expensive, both to create and maintain. On Tuesday we heard one estimate that custom software can cost as much as $100 per instruction. That is *per instruction,* and any useful software is likely to have thousands or tens of thousands of instructions. On the other hand, the presentations, demonstrations, and trade exhibits show us that for individual use most packaged software is almost free relative to the value that it has in it. If I can make a somewhat stretched analogy, the difference between custom software and purchased software is analogous to making your own movies versus buying a ticket and seeing one made by others. If at all possible, buy packaged software rather than write it.

9. Within the context of purchased software, other lessons emerge. While I would not say any exhibitor here has done it, outside of this Conference software is almost universally oversold. Claims are made for software that simply overstate what it actually can do. This is unfortunate, because most available software generally does quite a bit and there is little need to exaggerate. Nonetheless, a considerable amount of

exaggeration does go on. Consequently, you should bring a good deal of scepticism to the table when you evaluate packaged software. Bring the same scepticism that you do for any other business judgment. If you allow me another reference to my movie analogy, apply the same scepticism to software that you apply to the 'coming attractions' and promotions that are made for movies. On the other hand, do not agonise about a decision over packaged software. It is so relatively inexpensive that we can easily use up more time and money analysing it in depth than is involved in buying and using the software. For software for small computers, generally it makes sense to 'buy it and try it'. If you like it, great. If not, toss it out and look for something else.

10. What about the situation in which software does not exactly meet your application needs? I believe a major mistake we make is requiring software to meet 100 percent of our needs. Think for a moment about the total pounds that it will cost to satisfy 100 percent of requirements and visualise that money stacked in a pile. My experience suggests that roughly the lowest 10-20 percent of that pile (that is, about 10-20 percent of the total cost) is enough to meet 80-90 percent of application needs. The other 80-90 percent of the cost is spent to handle all the special cases and nuances that generally make up the final 10-20 percent of perceived needs. I believe it makes sense to be happy with getting 80-90 percent of our needs taken care of for 10-20 percent of the cost. Deal with the remaining 10-20 percent of needs in other ways. For example, it is not necessary to automate and computerise fully 100 percent of needs. Consider handling all or part of this 10-20 percent of needs in the 'old way'. Alternatively, make changes in the way these situations are treated so they fit the same mould as the other 80-90 percent.

Let me give an example from my own experience. Currently I am responsible for implementing a new order processing system to handle about 3 billion dollars of annual sales in the Weyerhaeuser Company. Developing software to handle 100 percent of the currently collected order data would cost well over $10 million, when considering software, hardware and related training. However, most of our order data can be handled by a software package plus hardware equipment that will cost only $5 million to implement. Rather than assume 100 percent of order data *must* be handled by the new system, we simply pointed out the fact that a small percent of data would cost us $5 million more to handle. Having the situation broken down this way allowed us to reach the obvious conclusions. We will not spend the second $5 million. We are modifying how we process sales so they can be processed by the basic $5 million system and we will put over $5 million in the bank.

Let us now address organisational lessons.

11. A brief word about required general purpose programming staff. A question might be: if we already have such people in organisation, should we be building up that staff; if we don't have any general purpose programming support now, is that an area of concern? I agree with Mark Pritchard's statement yesterday that the need for general purpose programming staff will not increase rapidly, and in fact may decline. This largely results from the trend towards relatively self-contained, easy-to-use software packages, and the parallel development of much better user-hardware interfaces. We have heard a good deal in this Conference about 'user friendly' systems. I believe that improvement of the interface between the user and the computer is going to be an area of continued major breakthroughs. This means that there is less need for general purpose programming 'middlemen'. This trend is already

evident at Weyerhaeuser Company. Even in computing companies such as ICL this trend is underway—as described by the speaker for ICL yesterday.

12. Another lesson to take from this Conference is that we should annually plan a small percentage of our organisation's budget for experimental computing. These funds should be set aside without an expectation of immediate specific returns. This money would be used to, for example, buy and try interesting forestry software that you see at a conference such as this, see advertised in the journals, or hear about somewhere else. You can use such money to explore a new computing approach to see whether it will be useful in your organisation. An example might involve a knowledge-based system such as described this morning. The experimemntal computing budget can be used to support attendance of state-of-the-art conferences, such as this one. This allows you and other members of your organisation to stay aware of what others are doing and how the trends are evolving. In this very fast changing field there are many new ideas and directions being thrust at us continually. A small, on-going expenditure for experimental computing helps us not to miss out on big opportunities and also helps to avoid spending big dollars on big mistakes. I am suggesting that you keep your toe in the water at all times, versus only occasionally diving in.

The final lesson I would raise is that we should treat computers in forestry as just one more professional and business variable—and deal with it accordingly. Do not give it special status. Too often, otherwise rational, hard-headed professionals and businessmen go soft-headed when it comes to computers. If you apply the same discipline, the same judgment, and the same intuition that you apply to every other aspect of your professional life, I believe you are assured of dealing with computers in forestry equally well.

Let me conclude by saying how much I have enjoyed being with you at this Conference this week. I am confident that for computers in forestry in the United Kingdom: *the best is yet to come.*

Thank you.